Foundations of Quantum Mechanics

Foundations of Quantum Mechanics

Foundations of Quantum Mechanics

Kendrick Porter

MURPHY & MOORE
www.murphy-moorepublishing.com

Foundations of Quantum Mechanics
Kendrick Porter
ISBN: 978-1-63987-238-1 (Hardback)

MURPHY & MOORE

Published by Murphy & Moore Publishing,
1 Rockefeller Plaza,
New York City, NY 10020, USA

Cataloging-in-Publication Data

Foundations of quantum mechanics / Kendrick Porter.
p. cm.
Includes bibliographical references and index.
ISBN 978-1-63987-238-1
1. Quantum theory. 2. Physics. I. Porter, Kendrick.
QC174.12 .P75 2022
530.12--dc23

For more information regarding Murphy & Moore Publishing and its products, please visit the publisher's website www.murphy-moorepublishing.com

Table of Contents

Preface VII

Chapter 1 Introduction 1
- Quantum Mechanics 1
- Matter Wave 27
- Uncertainity Principle 36

Chapter 2 The Wave Function 45
- Operator 66
- Mathematical Formulations 85
- Schrödinger Equation 97
- Expectation Values 125

Chapter 3 One-Dimensional Potentials 127
- Bound States 127
- One-Dimensional Infinite Potential Well 129
- One-Dimensional Finite Potential Well 136
- One-Dimensional Potential Barrier 143
- One-Dimensional Step Potential 149
- One-Dimensional Linear Harmonic Oscillator 153

Chapter 4 Three-Dimensional Potentials 156
- Three-Dimensional Box 156
- Spherically Symmetric Potentials 161
- Three-Dimensional Harmonic Oscillator 169

Chapter 5 Angular Momentum 170
- Angular Momentum Diagrams 171
- Orbital Angular Momentum 177

- Spin 180
- The Stern-Gerlach Experiment 194

Chapter 6 Theories and Methods 200
- Perturbation Theory 201
- Scattering Theory 219
- Scattering Length 221
- S-matrix 223

Permissions

Index

Preface

This book aims to help a broader range of students by exploring a wide variety of significant topics related to this discipline. It will help students in achieving a higher level of understanding of the subject and excel in their respective fields. This book would not have been possible without the unwavered support of my senior professors who took out the time to provide me feedback and help me with the process. I would also like to thank my family for their patience and support.

The theory of physics which delves into the physical properties of nature at an atomic and subatomic level is known as quantum mechanics. Quantum mechanics provides the foundations for quantum physics. This includes quantum field theory, quantum chemistry, quantum information science and quantum technology. This field seeks to explain such phenomena which cannot be understood by classical physics. The four most prominent phenomena which quantum mechanics aims to explain are principle of uncertainty, quantization of some physical properties, wave-particle duality and quantum entanglement. The diverse applications of this field include superconducting magnets, quantum optics, electron microscopy, light-emitting diodes, etc. This book is compiled in such a manner, that it will provide an in-depth knowledge about the theory and practice of this field. Such selected concepts that redefine quantum mechanics have been presented herein. Coherent flow of topics, student-friendly language and extensive use of examples make this book an invaluable source of knowledge.

A brief overview of the book contents is provided below:

Chapter - Introduction

Quantum mechanics is the branch of physics that deals with study of behavior and nature of the smallest atomic and subatomic particles. Matter waves, uncertainty principle, etc. are studied under its domain. This is an introductory chapter which will briefly introduce about quantum mechanics.

Chapter - The Wave Function

The wave function is defined as the variable quantity which describes the wave characteristics of a particle mathematically. Vector space structure, inner product, Hilbert space, Schrödinger equation, etc. are some of its concepts. The topics elaborated in this chapter will help in gaining a better perspective of wave function.

Chapter - One-Dimensional Potentials

One-Dimensional potentials include one-dimensional finite and infinite potential well, one-dimensional potential barrier, one-dimensional step potential, etc. This chapter has been carefully written to provide an easy understanding of the aspects related to one-dimensional potentials.

Chapter - Three-Dimensional Potentials

Three-Dimensional potentials include three-dimensional box, spherically symmetric potentials, three-dimensional harmonic oscillator, etc. The particle in a box model deals with the movement of free particles in space surrounded by impenetrable barriers. This chapter discusses all the concepts related to three-dimensional potentials in detail.

Chapter - Angular Momentum

Angular momentum is the product of inertia and angular velocity and defines the quantity of rotation of a body. Spin is a form of angular momentum which is carried by elementary and composite particles, and atomic nuclei. This chapter closely examines the concepts of angular momentum to provide an easy understanding of the subject.

Chapter - Theories and Methods

Approximation methods are useful for approximations to wave functions, energies, spectra, as well as transition probabilities and other dynamical quantities. Scattering theory deals with the studying of the scattering of waves and particles. This chapter sheds light on the methods and theories used in quantum mechanics for an in-depth understanding of the subject.

Kendrick Porter

1

Introduction

Quantum mechanics is the branch of physics that deals with study of behavior and nature of the smallest atomic and subatomic particles. Matter waves, uncertainty principle, etc. are studied under its domain. This is an introductory chapter which will briefly introduce about quantum mechanics.

QUANTUM MECHANICS

Quantum mechanics deals with the behaviour of matter and light on the atomic and subatomic scale. It attempts to describe and account for the properties of molecules and atoms and their constituents—electrons, protons, neutrons, and other more esoteric particles such as quarks and gluons. These properties include the interactions of the particles with one another and with electromagnetic radiation (i.e., light, X-rays, and gamma rays).

The behaviour of matter and radiation on the atomic scale often seems peculiar, and the consequences of quantum theory are accordingly difficult to understand and to believe. Its concepts frequently conflict with common-sense notions derived from observations of the everyday world. There is no reason, however, why the behaviour of the atomic world should conform to that of the familiar, large-scale world. It is important to realize that quantum mechanics is a branch of physics and that the business of physics is to describe and account for the way the world—on both the large and the small scale—actually is and not how one imagines it or would like it to be.

The study of quantum mechanics is rewarding for several reasons. First, it illustrates the essential methodology of physics. Second, it has been enormously successful in giving correct results in practically every situation to which it has been applied. There is, however, an intriguing paradox. In spite of the overwhelming practical success of quantum mechanics, the foundations of the subject contain unresolved problems—in particular, problems concerning the nature of measurement. An essential feature of quantum mechanics is that it is generally impossible, even in principle, to measure a system without disturbing it; the detailed nature of this disturbance and the exact point at which it occurs are obscure and controversial. Thus, quantum mechanics attracted some of the ablest scientists of the 20th century, and they erected what is perhaps the finest intellectual edifice of the period.

Basic Considerations

At a fundamental level, both radiation and matter have characteristics of particles and waves. The gradual recognition by scientists that radiation has particle-like properties and that matter has wavelike properties provided the impetus for the development of quantum mechanics. Influenced by Newton, most physicists of the 18th century believed that light consisted of particles, which they called corpuscles. From about 1800, evidence began to accumulate for a wave theory of light. At about this time Thomas Young showed that, if monochromatic light passes through a pair of slits, the two emerging beams interfere, so that a fringe pattern of alternately bright and dark bands appears on a screen. The bands are readily explained by a wave theory of light. According to the theory, a bright band is produced when the crests (and troughs) of the waves from the two slits arrive together at the screen; a dark band is produced when the crest of one wave arrives at the same time as the trough of the other, and the effects of the two light beams cancel. Beginning in 1815, a series of experiments by Augustin-Jean Fresnel of France and others showed that, when a parallel beam of light passes through a single slit, the emerging beam is no longer parallel but starts to diverge; this phenomenon is known as diffraction. Given the wavelength of the light and the geometry of the apparatus (i.e., the separation and widths of the slits and the distance from the slits to the screen), one can use the wave theory to calculate the expected pattern in each case; the theory agrees precisely with the experimental data.

Planck's Radiation Law

By the end of the 19th century, physicists almost universally accepted the wave theory of light. However, though the ideas of classical physics explain interference and diffraction phenomena relating to the propagation of light, they do not account for the absorption and emission of light. All bodies radiate electromagnetic energy as heat; in fact, a body emits radiation at all wavelengths. The energy radiated at different wavelengths is a maximum at a wavelength that depends on the temperature of the body; the hotter the body, the shorter the wavelength for maximum radiation. Attempts to calculate the energy distribution for the radiation from a blackbody using classical ideas were unsuccessful. (A blackbody is a hypothetical ideal body or surface that absorbs and reemits all radiant energy falling on it.) One formula, proposed by Wilhelm Wien of Germany, did not agree with observations at long wavelengths, and another, proposed by Lord Rayleigh of England, disagreed with those at short wavelengths.

In 1900 the German theoretical physicist Max Planck made a bold suggestion. He assumed that the radiation energy is emitted, not continuously, but rather in discrete packets called quanta. The energy E of the quantum is related to the frequency ν by E = hν. The quantity h, now known as Planck's constant, is a universal constant with the approximate value of 6.62607×10^{-34} joule·second. Planck showed that the calculated energy spectrum then agreed with observation over the entire wavelength range.

Einstein and the Photoelectric Effect

In 1905 Einstein extended Planck's hypothesis to explain the photoelectric effect, which is the emission of electrons by a metal surface when it is irradiated by light or more-energetic photons. The kinetic energy of the emitted electrons depends on the frequency ν of the radiation, not on its intensity; for a given metal, there is a threshold frequency ν_0 below which no electrons are emitted. Furthermore, emission takes place as soon as the light shines on the surface; there is no detectable delay. Einstein showed that these results can be explained by two assumptions: (1) that light is composed of corpuscles or photons, the energy of which is given by Planck's relationship, and (2) that an atom in the metal can absorb either a whole photon or nothing. Part of the energy of the absorbed photon frees an electron, which requires a fixed energy W, known as the work function of the metal; the rest is converted into the kinetic energy $m_e u^2/2$ of the emitted electron (m_e is the mass of the electron and u is its velocity). Thus, the energy relation is:

$$h\nu = W + \frac{m_e u^2}{2}.$$

If ν is less than ν_0, where $h\nu_0 = W$, no electrons are emitted. Not all the experimental results mentioned above were known in 1905, but all Einstein's predictions have been verified since.

Bohr's Theory of the Atom

A major contribution to the subject was made by Niels Bohr of Denmark, who applied the quantum hypothesis to atomic spectra in 1913. The spectra of light emitted by gaseous atoms had been studied extensively since the mid-19th century. It was found that radiation from gaseous atoms at low pressure consists of a set of discrete wavelengths. This is quite unlike the radiation from a solid, which is distributed over a continuous range of wavelengths. The set of discrete wavelengths from gaseous atoms is known as a line spectrum, because the radiation (light) emitted consists of a series of sharp lines. The wavelengths of the lines are characteristic of the element and may form extremely complex patterns. The simplest spectra are those of atomic hydrogen and the alkali atoms (e.g., lithium, sodium, and potassium). For hydrogen, the wavelengths λ are given by the empirical formula:

$$\frac{1}{\lambda} = R_\infty\left(\frac{1}{m^2} - \frac{1}{n^2}\right),$$

where m and n are positive integers with n > m and R_∞, known as the Rydberg constant, has the value 1.097373157 × 10^7 per metre. For a given value of m, the lines for varying n form a series. The lines for m = 1, the Lyman series, lie in the ultraviolet part of the spectrum; those for m = 2, the Balmer series, lie in the visible spectrum; and those for m = 3, the Paschen series, lie in the infrared.

Bohr started with a model suggested by the New Zealand-born British physicist Ernest Rutherford. The model was based on the experiments of Hans Geiger and Ernest Marsden, who in 1909 bombarded gold atoms with massive, fast-moving alpha particles; when some of these particles were deflected backward, Rutherford concluded that the atom has a massive, charged nucleus. In Rutherford's model, the atom resembles a miniature solar system with the nucleus acting as the Sun and the electrons as the circulating planets. Bohr made three assumptions. First, he postulated that, in contrast to classical mechanics, where an infinite number of orbits is possible, an electron can be in only one of a discrete set of orbits, which he termed stationary states. Second, he postulated that the only orbits allowed are those for which the angular momentum of the electron is a whole number n times ħ (ħ = h/2π). Third, Bohr assumed that Newton's laws of motion, so successful in calculating the paths of the planets around the Sun, also applied to electrons orbiting the nucleus. The force on the electron (the analogue of the gravitational force between the Sun and a planet) is the electrostatic attraction between the positively charged nucleus and the negatively charged electron. With these simple assumptions, he showed that the energy of the orbit has the form:

$$E_n = -\frac{E_0}{n^2},$$

where E_0 is a constant that may be expressed by a combination of the known constants e, m_e, and ħ. While in a stationary state, the atom does not give off energy as light; however, when an electron makes a transition from a state with energy E_n to one with lower energy E_m, a quantum of energy is radiated with frequency ν, given by the equation:

$$h\nu = E_n - E_m.$$

Inserting the expression for E_n into this equation and using the relation λν = c, where c is the speed of light, Bohr derived the formula for the wavelengths of the lines in the hydrogen spectrum, with the correct value of the Rydberg constant.

Bohr's theory was a brilliant step forward. Its two most important features have survived in present-day quantum mechanics. They are (1) the existence of stationary, non-radiating states and (2) the relationship of radiation frequency to the energy difference between the initial and final states in a transition. Prior to Bohr, physicists had thought that the radiation frequency would be the same as the electron's frequency of rotation in an orbit.

Scattering of X-rays

Soon scientists were faced with the fact that another form of radiation, X-rays, also exhibits both wave and particle properties. Max von Laue of Germany had shown in 1912 that crystals can be used as three-dimensional diffraction gratings for X-rays; his technique constituted the fundamental evidence for the wavelike nature of X-rays. The atoms of a crystal, which are arranged in a regular lattice, scatter the X-rays. For certain directions of scattering, all the crests of the X-rays coincide. (The scattered X-rays are said to be in phase and to give constructive interference.) For these directions, the scattered X-ray beam is very intense. Clearly, this phenomenon demonstrates wave behaviour. In fact, given the interatomic distances in the crystal and the directions of constructive interference, the wavelength of the waves can be calculated.

In 1922 the American physicist Arthur Holly Compton showed that X-rays scatter from electrons as if they are particles. Compton performed a series of experiments on the scattering of monochromatic, high-energy X-rays by graphite. He found that part of the scattered radiation had the same wavelength λ_o as the incident X-rays but that there was an additional component with a longer wavelength λ. To interpret his results, Compton regarded the X-ray photon as a particle that collides and bounces off an electron in the graphite target as though the photon and the electron were a pair of (dissimilar) billiard balls. Application of the laws of conservation of energy and momentum to the collision leads to a specific relation between the amount of energy transferred to the electron and the angle of scattering. For X-rays scattered through an angle θ, the wavelengths λ and λ_o are related by the equation:

$$\lambda - \lambda_0 = \frac{h}{m_e c}(1 - \cos\theta).$$

The experimental correctness of Compton's formula is direct evidence for the corpuscular behaviour of radiation.

De Broglie's Wave Hypothesis

Faced with evidence that electromagnetic radiation has both particle and wave characteristics, Louis-Victor de Broglie of France suggested a great unifying hypothesis in 1924. De Broglie proposed that matter has wave as well as particle properties. He suggested that material particles can behave as waves and that their wavelength λ is related to the linear momentum p of the particle by $\lambda = h/p$.

In 1927 Clinton Davisson and Lester Germer of the United States confirmed de Broglie's hypothesis for electrons. Using a crystal of nickel, they diffracted a beam of monoenergetic electrons and showed that the wavelength of the waves is related to the momentum of the electrons by the de Broglie equation. Since Davisson and Germer's investigation, similar experiments have been performed with atoms, molecules, neutrons, protons, and many other particles. All behave like waves with the same wavelength-momentum relationship.

Basic Concepts and Methods

Bohr's theory, which assumed that electrons moved in circular orbits, was extended by the German physicist Arnold Sommerfeld and others to include elliptic orbits and other refinements. Attempts were made to apply the theory to more complicated systems than the hydrogen atom. However, the ad hoc mixture of classical and quantum ideas made the theory and calculations increasingly unsatisfactory. Then, in the 12 months started in July 1925, a period of creativity without parallel in the history of physics, there appeared a series of papers by German scientists that set the subject on a firm conceptual foundation. The papers took two approaches: (1) matrix mechanics, proposed by Werner Heisenberg, Max Born, and Pascual Jordan, and (2) wave mechanics, put forward by Erwin Schrödinger. The protagonists were not always polite to each other. Heisenberg found the physical ideas of Schrödinger's theory "disgusting," and Schrödinger was "discouraged and repelled" by the lack of visualization in Heisenberg's method. However, Schrödinger, not allowing his emotions to interfere with his scientific endeavours, showed that, in spite of apparent dissimilarities, the two theories are equivalent mathematically.

Schrödinger's Wave Mechanics

Schrödinger expressed de Broglie's hypothesis concerning the wave behaviour of matter in a mathematical form that is adaptable to a variety of physical problems without additional arbitrary assumptions. He was guided by a mathematical formulation of optics, in which the straight-line propagation of light rays can be derived from wave motion when the wavelength is small compared to the dimensions of the apparatus employed. In the same way, Schrödinger set out to find a wave equation for matter that would give particle-like propagation when the wavelength becomes comparatively small. According to classical mechanics, if a particle of mass me is subjected to a force such that its potential energy is V(x, y, z) at position x, y, z, then the sum of V(x, y, z) and the kinetic energy $p^2/2m_e$ is equal to a constant, the total energy E of the particle. Thus,

$$\frac{p^2}{2m_e} + V(x,y,z) = E.$$

It is assumed that the particle is bound—i.e., confined by the potential to a certain region in space because its energy E is insufficient for it to escape. Since the potential varies with position, two other quantities do also: the momentum and, hence, by extension from the de Broglie relation, the wavelength of the wave. Postulating a wave function Ψ(x, y, z) that varies with position, Schrödinger replaced p in the above energy equation with a differential operator that embodied the de Broglie relation. He then showed that Ψ satisfies the partial differential equation:

$$-\frac{\hbar^2}{2m_e}\left(\frac{\partial^2\Psi}{\partial x^2}+\frac{\partial^2\Psi}{\partial y^2}+\frac{\partial^2\Psi}{\partial z^2}\right)+V(x,y,z)\Psi=E\Psi.$$

This is the (time-independent) Schrödinger wave equation, which established quantum mechanics in a widely applicable form. An important advantage of Schrödinger's theory is that no further arbitrary quantum conditions need be postulated. The required quantum results follow from certain reasonable restrictions placed on the wave function—for example, that it should not become infinitely large at large distances from the centre of the potential.

Schrödinger applied his equation to the hydrogen atom, for which the potential function, given by classical electrostatics, is proportional to $-e^2/r$, where −e is the charge on the electron. The nucleus (a proton of charge e) is situated at the origin, and r is the distance from the origin to the position of the electron. Schrödinger solved the equation for this particular potential with straightforward, though not elementary, mathematics. Only certain discrete values of E lead to acceptable functions Ψ. These functions are characterized by a trio of integers n, l, m, termed quantum numbers. The values of E depend only on the integers n (1, 2, 3, etc.) and are identical with those given by the Bohr theory. The quantum numbers l and m are related to the angular momentum of the electron; Square root of $\sqrt{l(l+1)}$ ħ is the magnitude of the angular momentum, and mħ is its component along some physical direction.

The square of the wave function, Ψ^2, has a physical interpretation. Schrödinger originally supposed that the electron was spread out in space and that its density at point x, y, z was given by the value of Ψ^2 at that point. Almost immediately Born proposed what is now the accepted interpretation—namely, that Ψ^2 gives the probability of finding the electron at x, y, z. The distinction between the two interpretations is important. If Ψ^2 is small at a particular position, the original interpretation implies that a small fraction of an electron will always be detected there. In Born's interpretation, nothing will be detected there most of the time, but, when something is observed, it will be a whole electron. Thus, the concept of the electron as a point particle moving in a well-defined path around the nucleus is replaced in wave mechanics by clouds that describe the probable locations of electrons in different states.

Electron Spin and Antiparticles

In 1928 the English physicist Paul A.M. Dirac produced a wave equation for the electron that combined relativity with quantum mechanics. Schrödinger's wave equation does not satisfy the requirements of the special theory of relativity because it is based on a nonrelativistic expression for the kinetic energy ($p^2/2m_e$). Dirac showed that an electron has an additional quantum number m_s. Unlike the first three quantum numbers,

m_s is not a whole integer and can have only the values $+^1/_2$ and $-^1/_2$. It corresponds to an additional form of angular momentum ascribed to a spinning motion. (The angular momentum mentioned above is due to the orbital motion of the electron, not its spin.) The concept of spin angular momentum was introduced in 1925 by Samuel A. Goudsmit and George E. Uhlenbeck, two graduate students at the University of Leiden, Neth., to explain the magnetic moment measurements made by Otto Stern and Walther Gerlach of Germany several years earlier. The magnetic moment of a particle is closely related to its angular momentum; if the angular momentum is zero, so is the magnetic moment. Yet Stern and Gerlach had observed a magnetic moment for electrons in silver atoms, which were known to have zero orbital angular momentum. Goudsmit and Uhlenbeck proposed that the observed magnetic moment was attributable to spin angular momentum.

The electron-spin hypothesis not only provided an explanation for the observed magnetic moment but also accounted for many other effects in atomic spectroscopy, including changes in spectral lines in the presence of a magnetic field (Zeeman effect), doublet lines in alkali spectra, and fine structure (close doublets and triplets) in the hydrogen spectrum.

The Dirac equation also predicted additional states of the electron that had not yet been observed. Experimental confirmation was provided in 1932 by the discovery of the positron by the American physicist Carl David Anderson. Every particle described by the Dirac equation has to have a corresponding antiparticle, which differs only in charge. The positron is just such an antiparticle of the negatively charged electron, having the same mass as the latter but a positive charge.

Identical Particles and Multielectron Atoms

Because electrons are identical to (i.e., indistinguishable from) each other, the wave function of an atom with more than one electron must satisfy special conditions. The problem of identical particles does not arise in classical physics, where the objects are large-scale and can always be distinguished, at least in principle. There is no way, however, to differentiate two electrons in the same atom, and the form of the wave function must reflect this fact. The overall wave function Ψ of a system of identical particles depends on the coordinates of all the particles. If the coordinates of two of the particles are interchanged, the wave function must remain unaltered or, at most, undergo a change of sign; the change of sign is permitted because it is Ψ^2 that occurs in the physical interpretation of the wave function. If the sign of Ψ remains unchanged, the wave function is said to be symmetric with respect to interchange; if the sign changes, the function is antisymmetric.

The symmetry of the wave function for identical particles is closely related to the spin of the particles. In quantum field theory, it can be shown that particles with half-integral spin ($^1/_2$, $^3/_2$, etc.) have antisymmetric wave functions. They are called fermions after the Italian-born physicist Enrico Fermi. Examples of fermions are electrons, protons, and neutrons, all of which have spin $^1/_2$. Particles with zero or integral spin (e.g., mesons, photons) have symmetric wave functions and are called bosons after the Indian mathematician and physicist Satyendra Nath Bose, who first applied the ideas of symmetry to photons in 1924–25.

The requirement of antisymmetric wave functions for fermions leads to a fundamental result, known as the exclusion principle, first proposed in 1925 by the Austrian physicist Wolfgang Pauli. The exclusion principle states that two fermions in the same system cannot be in the same quantum state. If they were, interchanging the two sets of coordinates would not change the wave function at all, which contradicts the result that the wave function must change sign. Thus, two electrons in the same atom cannot have an identical set of values for the four quantum numbers n, l, m, m_s. The exclusion principle forms the basis of many properties of matter, including the periodic classification of the elements, the nature of chemical bonds, and the behaviour of electrons in solids; the last determines in turn whether a solid is a metal, an insulator, or a semiconductor.

The Schrödinger equation cannot be solved precisely for atoms with more than one electron. The principles of the calculation are well understood, but the problems are complicated by the number of particles and the variety of forces involved. The forces include the electrostatic forces between the nucleus and the electrons and between the electrons themselves, as well as weaker magnetic forces arising from the spin and orbital motions of the electrons. Despite these difficulties, approximation methods introduced by the English physicist Douglas R. Hartree, the Russian physicist Vladimir Fock, and others in the 1920s and 1930s have achieved considerable success. Such schemes start by assuming that each electron moves independently in an average electric field because of the nucleus and the other electrons; i.e., correlations between the positions of the electrons are ignored. Each electron has its own wave function, called an orbital. The overall wave function for all the electrons in the atom satisfies the exclusion principle. Corrections to the calculated energies are then made, which depend on the strengths of the electron-electron correlations and the magnetic forces.

Time-dependent Schrödinger Equation

At the same time that Schrödinger proposed his time-independent equation to describe the stationary states, he also proposed a time-dependent equation to describe how a system changes from one state to another. By replacing the energy E in Schrödinger's equation with a time-derivative operator, he generalized his wave equation to determine the time variation of the wave function as well as its spatial variation. The time-dependent Schrödinger equation reads:

$$-\frac{\hbar h^2}{2m_e}\left(\frac{\partial^2\Psi}{\partial x^2}+\frac{\partial^2\Psi}{\partial y^2}+\frac{\partial^2\Psi}{\partial z^2}\right)+V(x,y,z)\Psi = i\hbar\frac{\partial\Psi}{\partial t}.$$

The quantity i is the square root of −1. The function Ψ varies with time t as well as with position x, y, z. For a system with constant energy, E, Ψ has the form:

$$\Psi(x,y,z,t)=\Psi(x,y,z)\exp\left(-\frac{iEt}{h\hbar}\right),$$

where exp stands for the exponential function, and the time-dependent Schrödinger equation reduces to the time-independent form.

The probability of a transition between one atomic stationary state and some other state can be calculated with the aid of the time-dependent Schrödinger equation. For example, an atommay change spontaneously from one state to another state with less energy, emitting the difference in energy as a photon with a frequency given by the Bohr relation. If electromagnetic radiation is applied to a set of atoms and if the frequency of the radiation matches the energy difference between two stationary states, transitions can be stimulated. In a stimulated transition, the energy of the atom may increase—i.e., the atom may absorb a photon from the radiation—or the energy of the atom may decrease, with the emission of a photon, which adds to the energy of the radiation. Such stimulated emission processes form the basic mechanism for the operation of lasers. The probability of a transition from one state to another depends on the values of the l, m, m_s quantum numbers of the initial and final states. For most values, the transition probability is effectively zero. However, for certain changes in the quantum numbers, summarized as selection rules, there is a finite probability. For example, according to one important selection rule, the l value changes by unity because photons have a spin of 1. The selection rules for radiation relate to the angular momentumproperties of the stationary states. The absorbed or emitted photon has its own angular momentum, and the selection rules reflect the conservation of angular momentum between the atoms and the radiation.

Tunneling

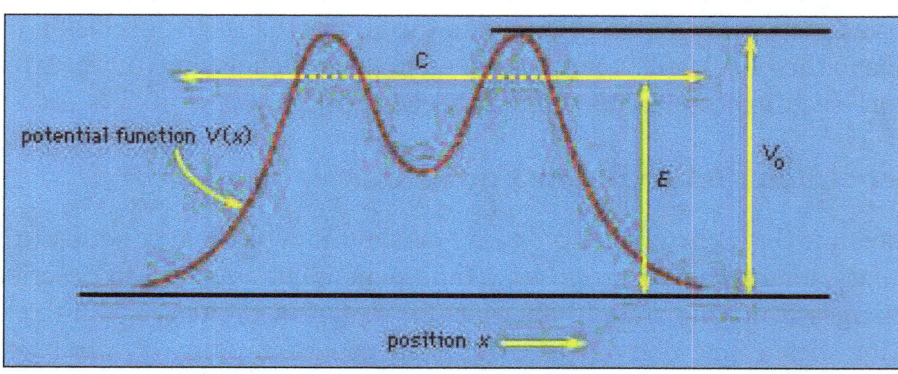

The phenomenon of tunneling. Classically, a particle is bound in the central region C if its energy E is less than Vo, but in quantum theory the particle may tunnel through the potential barrier and escape.

The phenomenon of tunneling, which has no counterpart in classical physics, is an important consequence of quantum mechanics. Consider a particle with energy E in the inner region of a one-dimensional potential well V(x), as shown in figure. (A potential well is a potential that has a lower value in a certain region of space than in the neighbouring regions.) In classical mechanics, if $E < V_0$ (the maximum height of the potential barrier), the particle remains in the well forever; if $E > V_0$, the particle escapes. In quantum mechanics, the situation is not so simple. The particle can escape even if its energy E is below the height of the barrier V_0, although the probability of escape is small unless E is close to V_0. In that case, the particle may tunnel through the potential barrier and emerge with the same energy E.

The phenomenon of tunneling has many important applications. For example, it describes a type of radioactive decay in which a nucleus emits an alpha particle (a helium nucleus). According to the quantum explanation given independently by George Gamow and by Ronald W. Gurney and Edward Condon in 1928, the alpha particle is confined before the decay by a potential For a given nuclear species, it is possible to measure the energy E of the emitted alpha particle and the average lifetime τ of the nucleus before decay. The lifetime of the nucleus is a measure of the probability of tunneling through the barrier—the shorter the lifetime, the higher the probability. With plausible assumptions about the general form of the potential function, it is possible to calculate a relationship between τ and E that is applicable to all alpha emitters. This theory, which is borne out by experiment, shows that the probability of tunneling, and hence the value of τ, is extremely sensitive to the value of E. For all known alpha-particle emitters, the value of E varies from about 2 to 8 million electron volts, or MeV (1 MeV = 10^6 electron volts). Thus, the value of E varies only by a factor of 4, whereas the range of τ is from about 10^{11} years down to about 10^{-6} second, a factor of 10^{24}. It would be difficult to account for this sensitivity of τ to the value of E by any theory other than quantum mechanical tunneling.

Axiomatic Approach

Although the two Schrödinger equations form an important part of quantum mechanics, it is possible to present the subject in a more general way. Dirac gave an elegant exposition of an axiomatic approach based on observables and states in a classic textbook entitled The Principles of Quantum Mechanics. An observable is anything that can be measured—energy, position, a component of angular momentum, and so forth. Every observable has a set of states, each state being represented by an algebraic function. With each state is associated a number that gives the result of a measurement of the observable. Consider an observable with N states, denoted by $\psi_1, \psi_2, \ldots, \psi_N$, and corresponding measurement values $a_1, a_2, \ldots, a_N$. A physical system—e.g., an atom in a particular state—is represented by a wave function Ψ, which can be expressed as a linear combination, or mixture, of the states of the observable. Thus, the Ψ may be written as:

$$\Psi = c_1\psi_1 + c_2\psi_2 + ... + c_N\psi_N.$$

For a given Ψ, the quantities c_1, c_2, etc., are a set of numbers that can be calculated. In general, the numbers are complex, but, in the present discussion, they are assumed to be real numbers.

The theory postulates, first, that the result of a measurement must be an a-value—i.e., a_1, a_2, or a_3, etc. No other value is possible. Second, before the measurement is made, the probability of obtaining the value a_1 is c_1^2, and that of obtaining the value a_2 is c_2^2, and so on. If the value obtained is, say, a_5, the theory asserts that after the measurement the state of the system is no longer the original Ψ but has changed to ψ_5, the state corresponding to a_5.

A number of consequences follow from these assertions. First, the result of a measurement cannot be predicted with certainty. Only the probability of a particular result can be predicted, even though the initial state (represented by the function Ψ) is known exactly. Second, identical measurements made on a large number of identical systems, all in the identical state Ψ, will produce different values for the measurements. This is, of course, quite contrary to classical physics and common sense, which say that the same measurement on the same object in the same state must produce the same result. Moreover, according to the theory, not only does the act of measurement change the state of the system, but it does so in an indeterminate way. Sometimes it changes the state to ψ_1, sometimes to ψ_2, and so forth.

There is an important exception to the above statements. Suppose that, before the measurement is made, the state Ψ happens to be one of the ψs—say, $\Psi = \psi_3$. Then $c_3 = 1$ and all the other cs are zero. This means that, before the measurement is made, the probability of obtaining the value a_3 is unity and the probability of obtaining any other value of a is zero. In other words, in this particular case, the result of the measurement can be predicted with certainty. Moreover, after the measurement is made, the state will be ψ_3, the same as it was before. Thus, in this particular case, measurement does not disturb the system. Whatever the initial state of the system, two measurements made in rapid succession (so that the change in the wave function given by the time-dependent Schrödinger equation is negligible) produce the same result.

The value of one observable can be determined by a single measurement. The value of two observables for a given system may be known at the same time, provided that the two observables have the same set of state functions $\psi_1, \psi_2, \ldots, \psi_N$. In this case, measuring the first observable results in a state function that is one of the ψs. Because this is also a state function of the second observable, the result of measuring the latter can be predicted with certainty. Thus the values of both observables are known. (Although the ψs are the same for the two observables, the two sets of a values are, in general, different). The two observables can be measured repeatedly in any sequence. After the first measurement, none of the measurements disturbs the system, and a unique pair of values for the two observables is obtained.

Incompatible Observables

The measurement of two observables with different sets of state functions is a quite different situation. Measurement of one observable gives a certain result. The state function after the measurement is, as always, one of the states of that observable; however, it is not a state function for the second observable. Measuring the second observable disturbs the system, and the state of the system is no longer one of the states of the first observable. In general, measuring the first observable again does not produce the same result as the first time. To sum up, both quantities cannot be known at the same time, and the two observables are said to be incompatible.

A specific example of this behaviour is the measurement of the component of angular momentum along two mutually perpendicular directions. The Stern-Gerlach experiment involved measuring the angular momentum of a silver atom in the ground state. In reconstructing this experiment, a beam of silver atoms is passed between the poles of a magnet. The poles are shaped so that the magnetic field varies greatly in strength over a very small distance. The apparatus determines the ms quantum number, which can be $+^1/_2$ or $-^1/_2$. No other values are obtained. Thus in this case the observable has only two states—i.e., N = 2. The inhomogeneous magnetic field produces a force on the silver atoms in a direction that depends on the spin state of the atoms. The result is shown schematically in figure. A beam of silver atoms is passed through magnet A. The atoms in the state with $m_s = +^1/_2$ are deflected upward and emerge as beam 1, while those with $m_s = -^1/_2$ are deflected downward and emerge as beam 2. If the direction of the magnetic field is the x-axis, the apparatus measures S_x, which is the x-component of spin angular momentum. The atoms in beam 1 have $S_x = +\hbar/2$ while those in beam 2 have $S_x = -\hbar/2$. In a classical picture, these two states represent atoms spinning about the direction of the x-axis with opposite senses of rotation.

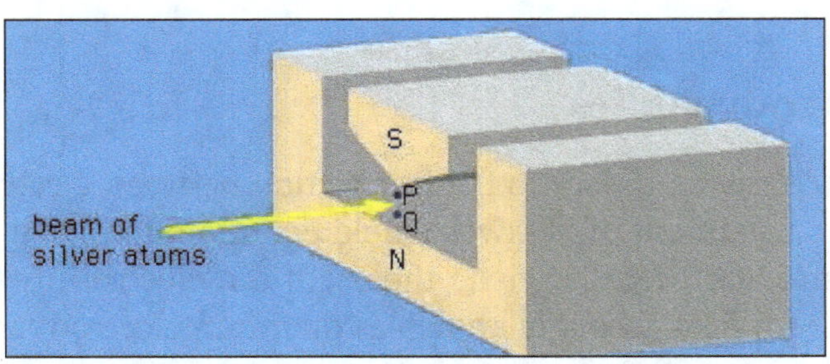

Magnet in Stern-Gerlach experiment. N and S are the north and south poles of a magnet. The knife-edge of S results in a much stronger magnetic field at the point P than at Q.

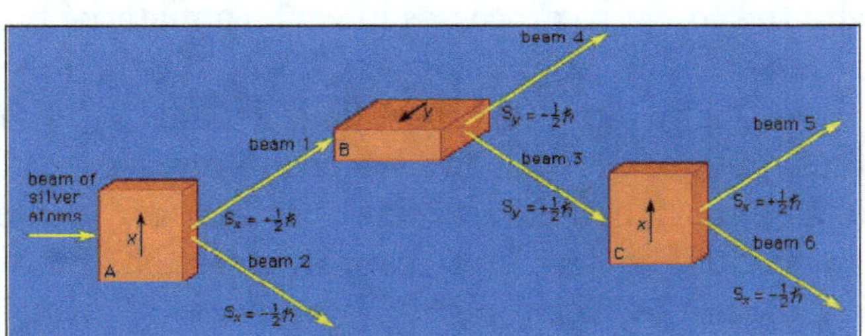

Measurements of the x and y components of angular momentum for silver atoms, S, in the ground state. A, B, and C are magnets with inhomogeneous magnetic fields. The arrows show the average direction of each magnetic field.

The y-component of spin angular momentum S_y also can have only the values $+\hbar/2$ and $-\hbar/2$; however, the two states of S_y are not the same as for S_x. In fact, each of the states of S_x is an equal mixture of the states for S_y, and conversely. Again, the two S_y states may be pictured as representing atoms with opposite senses of rotation about the y-axis. These classical pictures of quantum states are helpful, but only up to a certain point. For example, quantum theory says that each of the states corresponding to spin about the x-axis is a superposition of the two states with spin about the y-axis. There is no way to visualize this; it has absolutely no classical counterpart. One simply has to accept the result as a consequence of the axioms of the theory. Suppose that, as in figure, the atoms in beam 1 are passed into a second magnet B, which has a magnetic field along the y-axis perpendicular to x. The atoms emerge from B and go in equal numbers through its two output channels. Classical theory says that the two magnets together have measured both the x- and y-components of spin angular momentum and that the atoms in beam 3 have $S_x = +\hbar/2$, $S_y = +\hbar/2$, while those in beam 4 have $S_x = +\hbar/2$, $S_y = -\hbar/2$. However, classical theory is wrong, because if beam 3 is put through still another magnet C, with its magnetic field along x, the atoms divide equally into beams 5 and 6 instead of emerging as a single beam 5 (as they would if they had $S_x = +\hbar/2$). Thus, the correct statement is that the beam entering B has $S_x = +\hbar/2$ and is composed of an equal mixture of the states $S_y = +\hbar/2$ and $S_y = -\hbar/2$—i.e., the x-component of angular momentum is known but the y-component is not. Correspondingly, beam 3 leaving B has $S_y = +\hbar/2$ and is an equal mixture of the states $S_x = +\hbar/2$ and $S_x = -\hbar/2$; the y-component of angular momentum is known but the x-component is not. The information about S_x is lost because of the disturbance caused by magnet B in the measurement of S_y.

Quantum Electrodynamics

The application of quantum theory to the interaction between electrons and radiation requires a quantum treatment of Maxwell's field equations, which are the foundations of electromagnetism, and the relativistic theory of the electron formulated by Dirac. The resulting quantum field theory is known as quantum electrodynamics, or QED.

QED accounts for the behaviour and interactions of electrons, positrons, and photons. It deals with processes involving the creation of material particles from electromagnetic energy and with the converse processes in which a material particle and its antiparticle annihilate each other and produce energy. Initially the theory was beset with formidable mathematical difficulties, because the calculated values of quantities such as the charge and mass of the electron proved to be infinite. However, an ingenious set of techniques developed by Hans Bethe, Julian S. Schwinger, Tomonaga Shin'ichirō, Richard P. Feynman, and others dealt systematically with the infinities to obtain finite values of the physical quantities. Their method is known as renormalization. The theory has provided some remarkably accurate predictions.

According to the Dirac theory, two particular states in hydrogen with different quantum numbers have the same energy. QED, however, predicts a small difference in their energies; the difference may be determined by measuring the frequency of the electromagnetic radiation that produces transitions between the two states. This effect was first measured by Willis E. Lamb, Jr., and Robert Retherford in 1947. Its physical origin lies in the interaction of the electron with the random fluctuations in the surrounding electromagnetic field. These fluctuations, which exist even in the absence of an applied field, are a quantum phenomenon. The accuracy of experiment and theory in this area may be gauged by two recent values for the separation of the two states, expressed in terms of the frequency of the radiation that produces the transitions:

experiment $1{,}057{,}858 \pm 2$ kilohertz

theory $1{,}057{,}864 \pm 14$ kilohertz

An even more spectacular example of the success of QED is provided by the value for μ_e, the magnetic dipole moment of the free electron. Because the electron is spinning and has electric charge, it behaves like a tiny magnet, the strength of which is expressed by the value of μ_e. According to the Dirac theory, μ_e is exactly equal to $\mu_B = e\hbar/2m_e$, a quantity known as the Bohr magneton; however, QED predicts that $\mu_e = (1 + a)\mu_B$, where a is a small number, approximately $^1/_{860}$. Again, the physical origin of the QED correction is the interaction of the electron with random oscillations in the surrounding electromagnetic field. The best experimental determination of μ_e involves measuring not the quantity itself but the small correction term $\mu_e - \mu_B$. This greatly enhances the sensitivity of the experiment. The most recent results for the value of a are:

experiment $(115{,}965{,}219 \pm 1) \times 10^{-11}$

theory $(115{,}965{,}227 \pm 10) \times 10^{-11}$

Since *a* itself represents a small correction term, the magnetic dipole moment of the electron is measured with an accuracy of about one part in 10^{11}. One of the most precisely determined quantities in physics, the magnetic dipole moment of the electron can be calculated correctly from quantum theory to within about one part in 10^{10}.

Interpretation of Quantum Mechanics

Although quantum mechanics has been applied to problems in physics with great success, some of its ideas seem strange.

Electron: Wave or Particle

Young's aforementioned experiment in which a parallel beam of monochromatic light is passed through a pair of narrow parallel slits has an electron counterpart. In Young's original experiment, the intensity of the light varies with direction after passing through the slits. The intensity oscillates because of interference between the light waves emerging from the two slits, the rate of oscillation depending on the wavelength of the light and the separation of the slits. The oscillation creates a fringe pattern of alternating light and dark bands that is modulated by the diffraction pattern from each slit. If one of the slits is covered, the interference fringes disappear, and only the diffraction pattern is observed.

Young's experiment can be repeated with electrons all with the same momentum. The screen in the optical experiment is replaced by a closely spaced grid of electron detectors. There are many devices for detecting electrons; the most common are scintillators. When an electron passes through a scintillating material, such as sodium iodide, the material produces a light flash which gives a voltage pulse that can be amplified and recorded. The pattern of electrons recorded by each detector is the same as that predicted for waves with wavelengths given by the de Broglie formula. Thus, the experiment provides conclusive evidence for the wave behaviour of electrons.

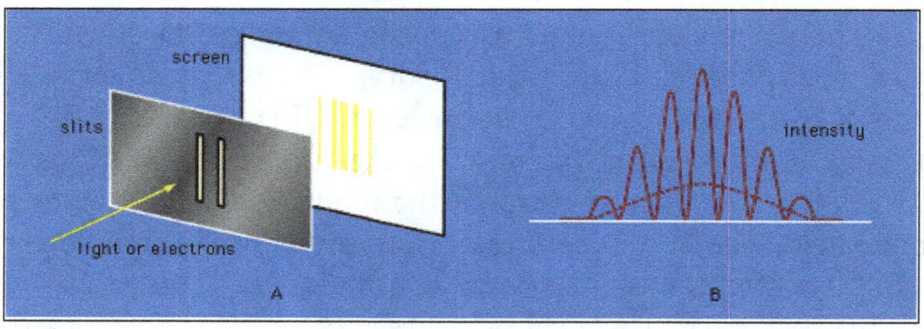

(A) Monochromatic light incident on a pair of slits gives interference fringes (alternate light and dark bands) on a screen, (B) variation in the intensity of the light at the screen when both slits are open. With a single slit, there is no interference pattern; the intensity variation is shown by the broken line.

If the experiment is repeated with a very weak source of electrons so that only one electron passes through the slits, a single detector registers the arrival of an electron. This is a well-localized event characteristic of a particle. Each time the experiment is repeated, one electron passes through the slits and is detected. A graph plotted with detector position along one axis and the number of electrons along the other looks exactly like the oscillating interference pattern in figure. Thus, the intensity function in the figure is proportional to the probability of the electron moving in a particular direction after it has passed through the slits. Apart from its units, the function is identical to Ψ^2, where Ψ is the solution of the time-independent Schrödinger equation for this particular experiment.

If one of the slits is covered, the fringe pattern disappears and is replaced by the diffraction pattern for a single slit. Thus, both slits are needed to produce the fringe pattern. However, if the electron is a particle, it seems reasonable to suppose that it passed through only one of the slits. The apparatus can be modified to ascertain which slit by placing a thin wire loop around each slit. When an electron passes through a loop, it generates a small electric signal, showing which slit it passed through. However, the interference fringe pattern then disappears, and the single-slit diffraction pattern returns. Since both slits are needed for the interference pattern to appear and since it is impossible to know which slit the electron passed through without destroying that pattern, one is forced to the conclusion that the electron goes through both slits at the same time.

The experiment shows both the wave and particle properties of the electron. The wave property predicts the probability of direction of travel before the electron is detected; on the other hand, the fact that the electron is detected in a particular place shows that it has particle properties. Therefore, the answer to the question whether the electron is a wave or a particle is that it is neither. It is an object exhibiting either wave or particle properties, depending on the type of measurement that is made on it. In other words, one cannot talk about the intrinsic properties of an electron; instead, one must consider the properties of the electron and measuring apparatus together.

Hidden Variables

A fundamental concept in quantum mechanics is that of randomness, or indeterminacy. In general, the theory predicts only the probability of a certain result. Consider the case of radioactivity. Imagine a box of atoms with identical nuclei that can undergo decay with the emission of an alpha particle. In a given time interval, a certain fraction will decay. The theory may tell precisely what that fraction will be, but it cannot predict which particular nuclei will decay. The theory asserts that, at the beginning of the time interval, all the nuclei are in an identical state and that the decay is a completely random process. Even in classical physics, many processes appear random. For example, one says that, when a roulette wheel is spun, the ball will drop at random into one of the numbered compartments in the wheel. Based on this belief, the casino owner and the players give and accept identical odds against each number for each throw. However, the fact is that the winning number could be predicted if one noted the exact location of the wheel when the croupier released the ball, the initial speed of the wheel, and various other physical parameters. It is only ignorance of the initial conditions and the difficulty of doing the calculations that makes the outcome appear to be random. In quantum mechanics, on the other hand, the randomness is asserted to be absolutely fundamental. The theory says that, though one nucleus decayed and the other did not, they were previously in the identical state.

Many eminent physicists, including Einstein, have not accepted this indeterminacy. They have rejected the notion that the nuclei were initially in the identical state. Instead, they postulated that there must be some other property—presently unknown, but existing nonetheless—that is different for the two nuclei. This type of unknown property is termed a hidden variable; if it existed, it would restore determinacy to physics. If the initial values of the hidden variables were known, it would be possible to predict which nuclei would decay. Such a theory would, of course, also have to account for the wealth of experimental data which conventional quantum mechanics explains from a few simple assumptions. Attempts have been made by de Broglie, David Bohm, and others to construct theories based on hidden variables, but the theories are very complicated and contrived. For example, the electron would definitely have to go through only one slit in the two-slit experiment. To explain that interference occurs only when the other slit is open, it is necessary to postulate a special force on the electron which exists only when that slit is open. Such artificial additions make hidden variable theories unattractive, and there is little support for them among physicists.

The orthodox view of quantum mechanics is known as the Copenhagen interpretation because its main protagonist, Niels Bohr, worked in that city. The Copenhagen view of understanding the physical world stresses the importance of basing theory on what can be observed and measured experimentally. It therefore rejects the idea of hidden variables as quantities that cannot be measured. The Copenhagen view is that the indeterminacy observed in nature is fundamental and does not reflect an inadequacy in present scientific knowledge. One should therefore accept the indeterminacy without trying to "explain" it and see what consequences come from it.

Attempts have been made to link the existence of free will with the indeterminacy of quantum mechanics, but it is difficult to see how this feature of the theory makes free will more plausible. On the contrary, free will presumably implies rational thought and decision, whereas the essence of the indeterminism in quantum mechanics is that it is due to intrinsic randomness.

Paradox of Einstein, Podolsky and Rosen

In 1935 Einstein and two other physicists in the United States, Boris Podolsky and Nathan Rosen, analyzed a thought experiment to measure position and momentum in a pair of interacting systems. Employing conventional quantum mechanics, they obtained some startling results, which led them to conclude that the theory does not give a complete description of physical reality. Their results, which are so peculiar as to seem paradoxical, are based on impeccable reasoning, but their conclusion that the theory is incomplete does not necessarily follow. Bohm simplified their experiment while retaining the central point of their reasoning; this discussion follows his account.

The proton, like the electron, has spin $^1/_2$; thus, no matter what direction is chosen for measuring the component of its spin angular momentum, the values are always $+\hbar/2$ or $-\hbar/2$. (The present discussion relates only to spin angular momentum, and the word spin is omitted from now on.) It is possible to obtain a system consisting of a pair of protons in close proximity and with total angular momentum equal to zero. Thus, if the value of one of the components of angular momentum for one of the protons is $+\hbar/2$ along any selected direction, the value for the component in the same direction for the other particle must be $-\hbar/2$. Suppose the two protons move in opposite directions until they are far apart. The total angular momentum of the system remains zero, and if the component of angular momentum along the same direction for each of the two particles is measured, the result is a pair of equal and opposite values. Therefore, after the quantity is measured for one of the protons, it can be predicted for the other proton; the second measurement is unnecessary. As previously noted, measuring a quantity changes the state of the system. Thus, if measuring S_x (the x-component of angular momentum) for proton 1 produces the value $+\hbar/2$, the state of proton 1 after measurement corresponds to $S_x = +\hbar/2$, and the state of proton 2 corresponds to $S_x = -\hbar/2$. Any direction, however, can be chosen for measuring the component of angular momentum. Whichever direction is selected, the state of proton 1 after measurement corresponds to a definite component of angular momentum about that direction. Furthermore, since proton 2 must have the opposite value for the same component, it follows that the measurement on proton 1 results in a definite state for proton 2 relative to the chosen direction, notwithstanding the fact that the two particles may be millions of kilometres apart and are not interacting with each other at the time. Einstein and his two collaborators thought that this conclusion was so obviously false that the quantum mechanical theory on which it was based must be incomplete. They concluded that the correct theory would contain some hidden variable feature that would restore the determinism of classical physics.

A comparison of how quantum theory and classical theory describe angular momentum for particle pairs illustrates the essential difference between the two outlooks. In both theories, if a system of two particles has a total angular momentum of zero, then the angular momenta of the two particles are equal and opposite. If the components of angular momentum are measured along the same direction, the two values are numerically equal, one positive and the other negative. Thus, if one component is measured, the other can be predicted. The crucial difference between the two theories is that, in classical physics, the system under investigation is assumed to have possessed the quantity being measured beforehand. The measurement does not disturb the system; it merely reveals the preexisting state. It may be noted that, if a particle were actually to possess components of angular momentum prior to measurement, such quantities would constitute hidden variables.

Does nature behave as quantum mechanics predicts? The answer comes from measuring the components of angular momenta for the two protons along different directions with an angle θ between them. A measurement on one proton can give only the result $+\hbar/2$ or $-\hbar/2$. The experiment consists of measuring correlations between the plus and minus values for pairs of protons with a fixed value of θ, and then repeating the measurements for different values of θ, as in figure. The interpretation of the results rests on an important theorem by the Irish-born physicist John Stewart Bell. Bell began by assuming the existence of some form of hidden variable with a value that would determine whether the measured angular momentum gives a plus or minus result. He further assumed locality—namely, that measurement on one proton (i.e., the choice of the measurement direction) cannot affect the result of the measurement on the other proton. Both these assumptions agree with classical, commonsense ideas. He then showed quite generally that these two assumptions lead to a certain relationship, now known as Bell's inequality, for the correlation values mentioned above. Experiments have been conducted at several laboratories with photons instead of protons (the analysis is similar), and the results show fairly conclusively that Bell's inequality is violated. That is to say, the observed results agree with those of quantum mechanics and cannot be accounted for by a hidden variable (or deterministic) theory based on the concept of locality. One is forced to conclude that the two protons are a correlated pair and that a measurement on one affects the state of both, no matter how far apart they are. This may strike one as highly peculiar, but such is the way nature appears to be.

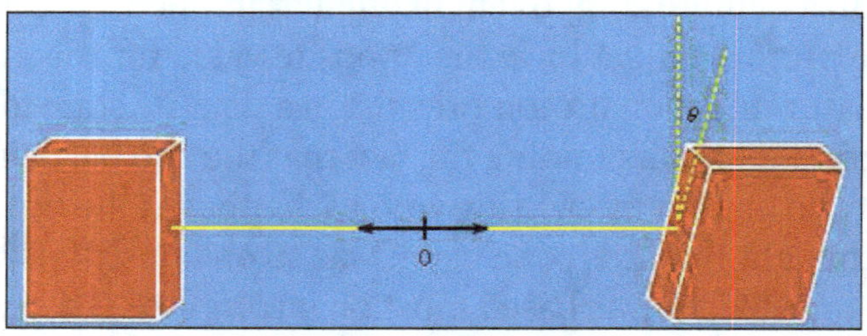

Experiment to determine the correlation in measured angular momentum values for a pair of protons with zero total angular momentum. The two protons are initially at the point o and move in opposite directions toward the two magnets.

It may be noted that the effect on the state of proton 2 following a measurement on proton 1 is believed to be instantaneous; the effect happens before a light signal initiated by the measuring event at proton 1 reaches proton 2. Alain Aspect and his coworkers in Paris demonstrated this result in 1982 with an ingenious experiment in which the correlation between the two angular momenta was measured, within a very short time interval, by a high-frequency switching device. The interval was less than the time taken for a light signal to travel from one particle to the other at the two measurement positions. Einstein's special theory of relativity states that no message can travel with a speed greater than that of light. Thus, there is no way that the information concerning the direction of the measurement on the first proton could reach the second proton before the measurement was made on it.

Measurement in Quantum Mechanics

The way quantum mechanics treats the process of measurement has caused considerable debate. Schrödinger's time-dependent wave equation (equation) is an exact recipe for determining the way the wave function varies with time for a given physical system in a given physical environment. According to the Schrödinger equation, the wave function varies in a strictly determinate way. On the other hand, in the axiomatic approach to quantum mechanics, a measurement changes the wave function abruptly and discontinuously. Before the measurement is made, the wave function Ψ is a mixture of the ψs as indicated in equation. The measurement changes Ψ from a mixture of ψs to a single ψ. This change, brought about by the process of measurement, is termed the collapse or reduction of the wave function. The collapse is a discontinuous change in Ψ; it is also unpredictable, because, starting with the same Ψ represented by the right-hand side of equation, the end result can be any one of the individual ψs.

The Schrödinger equation, which gives a smooth and predictable variation of Ψ, applies between the measurements. The measurement process itself, however, cannot be described by the Schrödinger equation; it is somehow a thing apart. This appears unsatisfactory, inasmuch as a measurement is a physical process and ought to be the subject of the Schrödinger equation just like any other physical process.

The difficulty is related to the fact that quantum mechanics applies to microscopic systems containing one (or a few) electrons, protons, or photons. Measurements, however, are made with large-scale objects (e.g., detectors, amplifiers, and meters) in the macroscopic world, which obeys the laws of classical physics. Thus, another way of formulating the question of what happens in a measurement is to ask how the microscopic quantum world relates and interacts with the macroscopic classical world. More narrowly, it can be asked how and at what point in the measurement process does the wave function collapse? So far, there are no satisfactory answers to these questions, although there are several schools of thought.

One approach stresses the role of a conscious observer in the measurement process and suggests that the wave function collapses when the observer reads the measuring instrument. Bringing the conscious mind into the measurement problem seems to raise more questions than it answers, however.

The Copenhagen interpretation of the measurement process is essentially pragmatic. It distinguishes between microscopic quantum systems and macroscopic measuring instruments. The initial object or event—e.g., the passage of an electron, photon, or atom—triggers the classical measuring device into giving a reading; somewhere along the chain of events, the result of the measurement becomes fixed (i.e., the wave function collapses). This does not answer the basic question but says, in effect, not to worry about it. This is probably the view of most practicing physicists.

A third school of thought notes that an essential feature of the measuring process is irreversibility. This contrasts with the behaviour of the wave function when it varies according to the Schrödinger equation; in principle, any such variation in the wave function can be reversed by an appropriate experimental arrangement. However, once a classical measuring instrument has given a reading, the process is not reversible. It is possible that the key to the nature of the measurement process lies somewhere here. The Schrödinger equation is known to apply only to relatively simple systems. It is an enormous extrapolation to assume that the same equation applies to the large and complex system of a classical measuring device. It may be that the appropriate equation for such a system has features that produce irreversible effects (e.g., wave-function collapse) which differ in kind from those for a simple system.

One may also mention the so-called many-worlds interpretation, proposed by Hugh Everett III in 1957, which suggests that, when a measurement is made for a system in which the wave function is a mixture of states, the universe branches into a number of noninteracting universes. Each of the possible outcomes of the measurement occurs, but in a different universe. Thus, $S_x = \frac{1}{2}$ is the result of a Stern-Gerlach measurement on a silver atom, there is another universe identical to ours in every way (including clones of people), except that the result of the measurement is Sx = $-^1/_2$. Although this fanciful model solves some measurement problems, it has few adherents among physicists.

Because the various ways of looking at the measurement process lead to the same experimental consequences, trying to distinguish between them on scientific grounds may be fruitless. One or another may be preferred on the grounds of plausibility, elegance, or economy of hypotheses, but these are matters of individual taste. Whether one day a satisfactory quantum theory of measurement will emerge, distinguished from the others by its verifiable predictions, remains an open question.

Applications of Quantum Mechanics

Quantum mechanics has been enormously successful in explaining microscopic phenomena in all branches of physics. The three phenomena described in this topic are examples that demonstrate the quintessence of the theory.

Decay of the Kaon

The kaon (also called the K^0 meson), discovered in 1947, is produced in high-energy collisions between nuclei and other particles. It has zero electric charge, and its mass is about one-half the mass of the proton. It is unstable and, once formed, rapidly decays into either 2 or 3 pi-mesons. The average lifetime of the kaon is about 10^{-10} second.

In spite of the fact that the kaon is uncharged, quantum theory predicts the existence of an antiparticle with the same mass, decay products, and average lifetime; the antiparticle is denoted by $\bar{K}^0$. During the early 1950s, several physicists questioned the justification for postulating the existence of two particles with such similar properties. In 1955, however, Murray Gell-Mann and Abraham Pais made an interesting prediction about the decay of the kaon. Their reasoning provides an excellent illustration of the quantum mechanical axiom that the wave function Ψ can be a superposition of states; in this case, there are two states, the K^0 and $\bar{K}^0$ mesons themselves.

A K^0 meson may be represented formally by writing the wave function as $\Psi = K^0$;

similarly $\Psi = \overline{K}^o$ represents a $\overline{K}^o$ meson. From the two states, K^o and $\overline{K}^o$, the following two new states are constructed:

$$K_1 = \frac{\left(K^0 + \overline{K}^0\right)}{\sqrt{2}},$$

$$K_2 = \frac{\left(K^0 - \overline{K}^0\right)}{\sqrt{2}}.$$

From these two equations it follows that:

$$K^0 = \frac{\left(K_1 + K_2\right)}{\sqrt{2}},$$

$$\overline{K}^0 = \frac{\left(K_1 - K_2\right)}{\sqrt{2}}.$$

The reason for defining the two states K_1 and K_2 is that, according to quantum theory, when the K^o decays, it does not do so as an isolated particle; instead, it combines with its antiparticle to form the states K_1 and K_2. The state K_1 (called the K-short [K^o_S]) decays into two pi-mesonswith a very short lifetime (about 9×10^{-11} second), while K_2 (called the K-long [K^o_L]) decays into three pi-mesons with a longer lifetime (about 5×10^{-8} second).

The physical consequences of these results may be demonstrated in the following experiment. K^o particles are produced in a nuclear reaction at the point A. They move to the right in the figure and start to decay. At point A, the wave function is $\Psi = K^o$, which, from equation $K^0 = \frac{\left(K_1 + K_2\right)}{\sqrt{2}}$, can be expressed as the sum of K_1 and K_2. As the particles move to the right, the K_1 state begins to decay rapidly. If the particles reach point B in about 10^{-8} second, nearly all the K_1component has decayed, although hardly any of the K_2 component has done so. Thus, at point B, the beam has changed from one of pure K^o to one of almost pure K_2, which equation $K_2 = \frac{\left(K^0 - \overline{K}^0\right)}{\sqrt{2}}$ shows is an equal mixture of K^o and $\overline{K}^o$. In other words, $\overline{K}^o$ particles appear in the beam simply because K_1 and K_2 decay at different rates. At point B, the beam enters a block of absorbing material. Both the K^o and $\overline{K}^o$ are absorbed by the nuclei in the block, but the $\overline{K}^o$ are absorbed more strongly. As a result, even though the beam is an equal mixture of K^o and $\overline{K}^o$ when it enters the absorber, it is almost pure K^o when it exits at point C. The beam thus begins and ends as K^o.

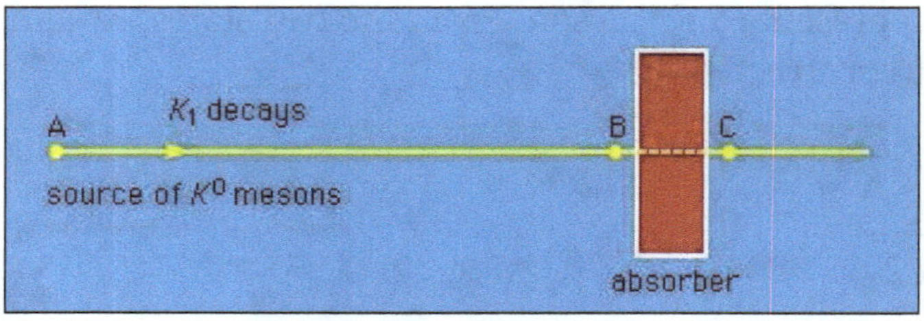

Decay of the K° meson.

Gell-Mann and Pais predicted all this, and experiments subsequently verified it. The experimental observations are that the decay products are primarily two pi-mesons with a short decay time near A, three pi-mesons with longer decay time near B, and two pi-mesons again near C. (This account exaggerates the changes in the K_1 and K_2 components between A and B and in the K° and $\bar{K}$° components between B and C; the argument, however, is unchanged.) The phenomenon of generating the $\bar{K}$° and regenerating the K_1 decay is purely quantum. It rests on the quantum axiom of the superposition of states and has no classical counterpart.

Cesium Clock

The cesium clock is the most accurate type of clock yet developed. This device makes use of transitions between the spin states of the cesium nucleus and produces a frequency which is so regular that it has been adopted for establishing the time standard.

Like electrons, many atomic nuclei have spin. The spin of these nuclei produces a set of small effects in the spectra, known as hyperfine structure. (The effects are small because, though the angular momentum of a spinning nucleus is of the same magnitude as that of an electron, its magnetic moment, which governs the energies of the atomic levels, is relatively small.) The nucleus of the cesium atom has spin quantum number $^7/_2$. The total angular momentum of the lowest energy states of the cesium atom is obtained by combining the spin angular momentum of the nucleus with that of the single valence electron in the atom. (Only the valence electron contributes to the angular momentum because the angular momenta of all the other electrons total zero. Another simplifying feature is that the ground states have zero orbital momenta, so only spin angular momenta need to be considered.) When nuclear spin is taken into account, the total angular momentum of the atom is characterized by a quantumnumber, conventionally denoted by F, which for cesium is 4 or 3. These values come from the spin value $^7/_2$ for the nucleus and $^1/_2$ for the electron. If the nucleus and the electron are visualized as tiny spinning tops, the value F = 4 ($^7/_2$ + $^1/_2$) corresponds to the tops spinning in the same sense, and F = 3 ($^7/_2$ − $^1/_2$) corresponds to spins in opposite senses. The energy difference ΔE of the states with the two F values is a precise quantity. If electromagnetic radiation of frequency ν_0, where:

$$h\nu_0 = \Delta E,$$

is applied to a system of cesium atoms, transitions will occur between the two states. An apparatus that can detect the occurrence of transitions thus provides an extremely precise frequency standard. This is the principle of the cesium clock.

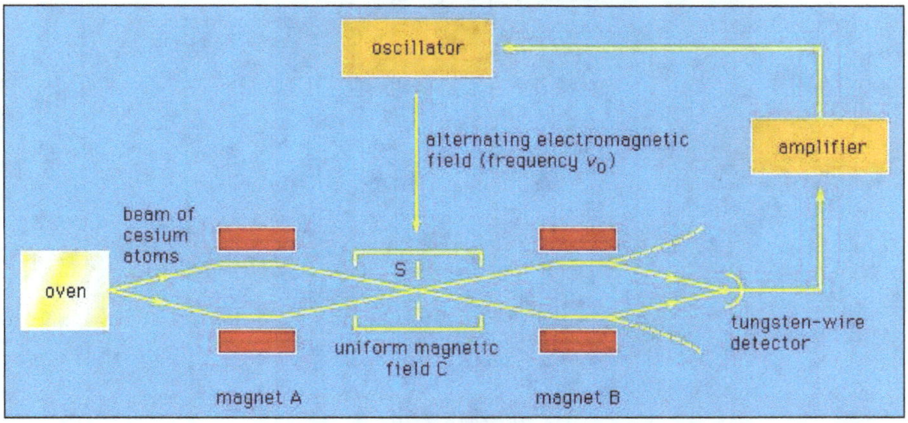

Cesium clock.

The apparatus is shown schematically in figure. A beam of cesium atoms emerges from an oven at a temperature of about 100 °C. The atoms pass through an inhomogeneous magnet A, which deflects the atoms in state F = 4 downward and those in state F = 3 by an equal amount upward. The atoms pass through slit S and continue into a second inhomogeneous magnet B. Magnet B is arranged so that it deflects atoms with an unchanged state in the same direction that magnet A deflected them. The atoms follow the paths indicated by the broken lines in the figure and are lost to the beam. However, if an alternating electromagnetic field of frequency ν_0 is applied to the beam as it traverses the centre region C, transitions between states will occur. Some atoms in state F = 4 will change to F = 3, and vice versa. For such atoms, the deflections in magnet B are reversed. The atoms follow the whole lines in the diagram and strike a tungsten wire, which gives electric signals in proportion to the number of cesium atoms striking the wire. As the frequency ν of the alternating field is varied, the signal has a sharp maximum for $\nu = \nu_0$. The length of the apparatus from the oven to the tungsten detector is about one metre.

Each atomic state is characterized not only by the quantum number F but also by a second quantum number m_F. For F = 4, m_F can take integral values from 4 to −4. In the absence of a magnetic field, these states have the same energy. A magnetic field, however, causes a small change in energy proportional to the magnitude of the field and to the m_F value. Similarly, a magnetic field changes the energy for the F = 3 states according to the m_F value which, in this case, may vary from 3 to −3. The energy changes are indicated in figure. In the cesium clock, a weak constant magnetic field is superposed on the alternating electromagnetic field in region C. The theory shows that the alternating field can bring about a transition only between pairs of states with m_F values that are the same or that differ by unity. However, as can be seen from the figure, the only transitions occurring at the frequency ν_0 are those between the two states with $m_F = 0$. The apparatus is so sensitive that it can discriminate easily between such transitions and all the others.

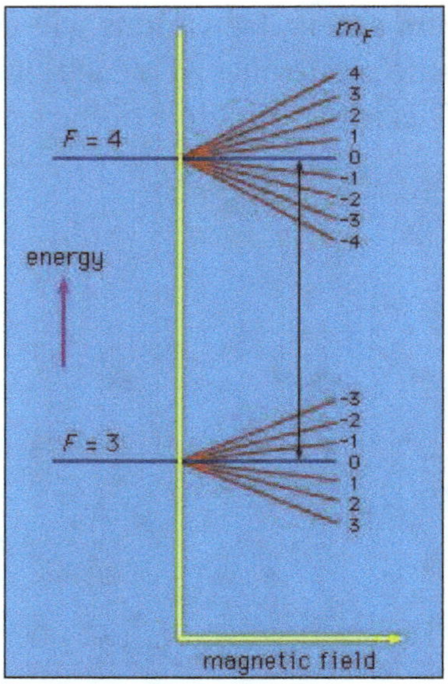

Variation of energy with magnetic-field strength for the F = 4 and F = 3 states in cesium-133.

If the frequency of the oscillator drifts slightly so that it does not quite equal ν_0, the detector output drops. The change in signal strength produces a signal to the oscillator to bring the frequency back to the correct value. This feedback system keeps the oscillator frequency automatically locked to ν_0.

The cesium clock is exceedingly stable. The frequency of the oscillator remains constant to about one part in 10^{13}. For this reason, the device is used to redefine the second. This base unit of time in the SI system is defined as equal to 9,192,631,770 cycles of the radiation corresponding to the transition between the levels F = 4, m_F = 0 and F = 3, m_F = 0 of the ground state of the cesium-133 atom. Prior to 1967, the second was defined in terms of the motion of Earth. The latter, however, is not nearly as stable as the cesium clock. Specifically, the fractional variation of Earth's rotation period is a few hundred times larger than that of the frequency of the cesium clock.

Quantum Voltage Standard

Quantum theory has been used to establish a voltage standard, and this standard has proven to be extraordinarily accurate and consistent from laboratory to laboratory.

If two layers of superconducting material are separated by a thin insulating barrier, a supercurrent (i.e., a current of paired electrons) can pass from one superconductor to the other. This is another example of the tunneling process described earlier. Several effects based on this phenomenon were predicted in 1962 by the British physicist Brian D. Josephson. Demonstrated experimentally soon afterwards, they are now referred to as the Josephson effects.

If a DC (direct-current) voltage V is applied across the two superconductors, the energy of an electron pair changes by an amount of 2eV as it crosses the junction. As a result, the supercurrent oscillates with frequency ν given by the Planck relationship (E = hν). Thus,

$$2eV = h\nu$$

This oscillatory behaviour of the supercurrent is known as the AC (alternating-current) Josephson effect. Measurement of V and ν permits a direct verification of the Planck relationship. Although the oscillating supercurrent has been detected directly, it is extremely weak. A more sensitive method of investigating in the above equation is to study effects resulting from the interaction of microwave radiation with the supercurrent.

Several carefully conducted experiments have verified in the above equation to such a high degree of precision that it has been used to determine the value of 2e/h. This value can in fact be determined more precisely by the AC Josephson effect than by any other method. The result is so reliable that laboratories now employ the AC Josephson effect to set a voltage standard. The numerical relationship between V and ν is:

$$\frac{2e}{h} = \frac{\nu}{h} = 482{,}597.9 \times 10^{9} \text{ hertz per volt}$$

In this way, measuring a frequency, which can be done with great precision, gives the value of the voltage. Before the Josephson method was used, the voltage standard in metrological laboratories devoted to the maintenance of physical units was based on high-stability Weston cadmium cells. These cells, however, tend to drift and so caused inconsistencies between standards in different laboratories. The Josephson method has provided a standard giving agreement to within a few parts in 10^8 for measurements made at different times and in different laboratories.

The values of the fundamental constants, such as c, h, e, and m_e, are determined from a wide variety of experiments based on quantum phenomena. The results are so consistent that the values of the constants are thought to be known in most cases to better than one part in 10^6. Physicists may not know what they are doing when they make a measurement, but they do it extremely well.

MATTER WAVE

Matter waves are a central part of the theory of quantum mechanics, being an example of wave–particle duality. All matter can exhibit wave-like behavior. For example, a beam of electrons can be diffracted just like a beam of light or a water wave. The concept that matter behaves like a wave was proposed by Louis de Broglie in 1924. It is also referred to as the de Broglie hypothesis. Matter waves are referred to as de Broglie waves.

The de Broglie wavelength is the wavelength, λ, associated with a massive particle and is related to its momentum, p, through the Planck constant, h:

$$\lambda = \frac{h}{p} = \frac{h}{mv}.$$

Wave-like behavior of matter was first experimentally demonstrated by George Paget Thomson's thin metal diffraction experiment, and independently in the Davisson–Germer experiment both using electrons, and it has also been confirmed for other elementary particles, neutral atoms and even molecules.

De Broglie Hypothesis

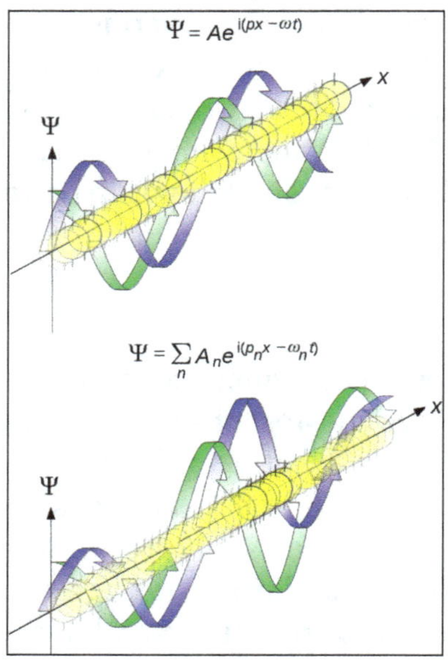

Propagation of de Broglie waves in 1d – real part of the complex amplitude is blue, imaginary part is green. The probability (shown as the color opacity) of finding the particle at a given point x is spread out like a waveform; there is no definite position of the particle. As the amplitude increases above zero the curvature decreases, so the amplitude decreases again, and vice versa. The result is an alternating amplitude: a wave. Top: plane wave. Bottom: wave packet.

De Broglie, in his 1924 PhD thesis, proposed that just as light has both wave-like and particle-like properties, electrons also have wave-like properties. By rearranging the momentum equation stated in the above topic, we find a relationship between the wavelength, λ associated with an electron and its momentum, p, through the Planck constant, h:

$$\lambda = \frac{h}{p}.$$

The relationship is now known to hold for all types of matter: all matter exhibits properties of both particles and waves.

> "When I conceived the first basic ideas of wave mechanics in 1923–24, I was guided by the aim to perform a real physical synthesis, valid for all particles, of the coexistence of the wave and of the corpuscular aspects that Einstein had introduced for photons in his theory of light quanta in 1905."
>
> — De Broglie

In 1926, Erwin Schrödinger published an equation describing how a matter wave should evolve—the matter wave analogue of Maxwell's equations—and used it to derive the energy spectrum of hydrogen.

Experimental Confirmation

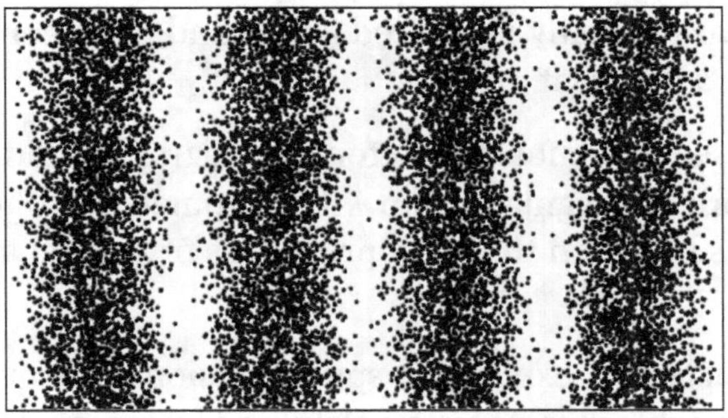

Demonstration of a matter wave in diffraction of electrons.

Matter waves were first experimentally confirmed to occur in George Paget Thomson's cathode ray diffraction experiment and the Davisson-Germer experiment for electrons, and the de Broglie hypothesis has been confirmed for other elementary particles. Furthermore, neutral atoms and even molecules have been shown to be wave-like.

Electrons

In 1927 at Bell Labs, Clinton Davisson and Lester Germer fired slow-moving electrons at a crystalline nickel target. The angular dependence of the diffracted electron intensity was measured, and was determined to have the same diffraction pattern as those predicted by Bragg for x-rays. At the same time George Paget Thomson at the University of Aberdeen was independently firing electrons at very thin metal foils to demonstrate the same effect. Before the acceptance of the de Broglie hypothesis, diffraction was a property that was thought to be exhibited only by waves. Therefore, the presence of any diffraction effects by matter demonstrated the wave-like nature of matter. When the de Broglie wavelength was inserted into the Bragg condition, the observed diffraction pattern was predicted, thereby experimentally confirming the de Broglie hypothesis for electrons.

This was a pivotal result in the development of quantum mechanics. Just as the photoelectric effect demonstrated the particle nature of light, the Davisson–Germer experiment showed the wave-nature of matter, and completed the theory of wave–particle duality. For physicists this idea was important because it meant that not only could any particle exhibit wave characteristics, but that one could use wave equations to describe phenomena in matter if one used the de Broglie wavelength.

Neutral Atoms

Experiments with Fresnel diffraction and an atomic mirror for specular reflection of neutral atoms confirm the application of the de Broglie hypothesis to atoms, i.e. the existence of atomic waves which undergo diffraction, interference and allow quantum reflection by the tails of the attractive potential. Advances in laser cooling have allowed cooling of neutral atoms down to nanokelvin temperatures. At these temperatures, the thermal de Broglie wavelengths come into the micrometre range. Using Bragg diffraction of atoms and a Ramsey interferometry technique, the de Broglie wavelength of cold sodium atoms was explicitly measured and found to be consistent with the temperature measured by a different method.

This effect has been used to demonstrate atomic holography, and it may allow the construction of an atom probe imaging system with nanometer resolution. The description of these phenomena is based on the wave properties of neutral atoms, confirming the de Broglie hypothesis.

The effect has also been used to explain the spatial version of the quantum Zeno effect, in which an otherwise unstable object may be stabilised by rapidly repeated observations.

Molecules

Recent experiments even confirm the relations for molecules and even macromolecules that otherwise might be supposed too large to undergo quantum mechanical effects. In 1999, a research team in Vienna demonstrated diffraction for molecules as large as fullerenes. The researchers calculated a De Broglie wavelength of the most probable C_{60} velocity as 2.5 pm. More recent experiments prove the quantum nature of molecules made of 810 atoms and with a mass of 10,123 amu.

Still one step further than Louis de Broglie go theories which in quantum mechanics eliminate the concept of a pointlike classical particle and explain the observed facts by means of wavepackets of matter waves alone.

De Broglie Relations

The de Broglie equations relate the wavelength λ to the momentum p, and frequency f to the total energy E of a particle:

$$\lambda = h/p$$
$$f = E/h$$

where h is the Planck constant. The equations can also be written as:

$$p = \hbar k$$
$$E = \hbar \omega$$

Or,

$$p = \hbar \beta$$
$$E = \hbar \omega$$

where $\hbar = h/2\pi$ is the reduced Planck constant, k is the wave vector, β is the phase constant, and ω is the angular frequency.

In each pair, the second equation is also referred to as the Planck–Einstein relation, since it was also proposed by Planck and Einstein.

Special Relativity

Using two formulas from special relativity, one for the relativistic momentum and one for the relativistic mass energy:

$$E = mc^2 = \gamma m_0 c^2$$
$$\vec{p} = m\vec{v} = \gamma m_0 \vec{v}$$

allows the equations to be written as:

$$\lambda = \frac{h}{\gamma m_0 v} = \frac{h}{m_0 v}\sqrt{1 - \frac{v^2}{c^2}}$$

$$f = \frac{c}{\lambda} = \frac{\gamma m_0 v c}{h} = \frac{m_0 c^2}{h} / \sqrt{\frac{c^2}{v^2} - 1}$$

where m_0 denotes the particle's rest mass, v its velocity, γ the Lorentz factor, and c the speed of light in a vacuum. Group velocity (equal to the particle's speed) should not be confused with phase velocity (equal to the product of the particle's frequency and its wavelength). In the case of a non-dispersive medium, they happen to be equal, but otherwise they are not.

Group Velocity

Albert Einstein first explained the wave–particle duality of light in 1905. Louis de Broglie hypothesized that any particle should also exhibit such a duality. The velocity of a particle, he concluded, should always equal the group velocity of the corresponding wave. The magnitude of the group velocity is equal to the particle's speed.

Both in relativistic and non-relativistic quantum physics, we can identify the group velocity of a particle's wave function with the particle velocity. Quantum mechanics has very accurately demonstrated this hypothesis, and the relation has been shown explicitly for particles as large as molecules.

De Broglie deduced that if the duality equations already known for light were the same for any particle, then his hypothesis would hold. This means that:

$$v_g = \frac{\partial \omega}{\partial k} = \frac{\partial (E/\hbar)}{\partial (p/\hbar)} = \frac{\partial E}{\partial p}$$

where E is the total energy of the particle, p is its momentum, ħ is the reduced Planck constant. For a free non-relativistic particle it follows that:

$$\begin{aligned} v_g &= \frac{\partial E}{\partial p} = \frac{\partial}{\partial p}\left(\frac{1}{2}\frac{p^2}{m}\right) \\ &= \frac{p}{m} \\ &= v \end{aligned}$$

where m is the mass of the particle and v its velocity.

Also in special relativity we find that,

$$\begin{aligned} v_g &= \frac{\partial E}{\partial p} = \frac{\partial}{\partial p}\left(\sqrt{p^2c^2 + m_0^2c^4}\right) \\ &= \frac{pc^2}{\sqrt{p^2c^2 + m_0^2c^4}} \\ &= \frac{pc^2}{E} \end{aligned}$$

where m_0 is the rest mass of the particle and c is the speed of light in a vacuum. But using that the phase velocity is $v_p = E/p = c^2/v$, therefore:

$$\begin{aligned} v_g &= \frac{pc^2}{E} \\ &= \frac{c^2}{v_p} \\ &= v \end{aligned}$$

where v is the velocity of the particle regardless of wave behavior.

Phase Velocity

In quantum mechanics, particles also behave as waves with complex phases. The phase velocity is equal to the product of the frequency multiplied by the wavelength.

By the de Broglie hypothesis, we see that:

$$v_{\mathrm{p}} = \frac{\omega}{k} = \frac{E/\hbar}{p/\hbar} = \frac{E}{p}$$

Using relativistic relations for energy and momentum, we have:

$$v_{\mathrm{p}} = \frac{E}{p} = \frac{mc^2}{mv} = \frac{\gamma m_0 c^2}{\gamma m_0 v} = \frac{c^2}{v} = \frac{c}{\beta}$$

where E is the total energy of the particle (i.e. rest energy plus kinetic energy in the kinematic sense), p the momentum, γ the Lorentz factor, c the speed of light, and β the speed as a fraction of c. The variable v can either be taken to be the speed of the particle or the group velocity of the corresponding matter wave. Since the particle speed $v < c$ for any particle that has mass (according to special relativity), the phase velocity of matter waves always exceeds c, i.e.:

$$v_{\mathrm{p}} > c,$$

and as we can see, it approaches c when the particle speed is in the relativistic range. The superluminal phase velocity does not violate special relativity, because phase propagation carries no energy.

Four-vectors

Using four-vectors, the De Broglie relations form a single equation:

$$\mathbf{P} = \hbar\mathbf{K}$$

which is frame-independent.

Likewise, the relation between group/particle velocity and phase velocity is given in frame-independent form by:

$$\mathbf{K} = \left(\frac{\omega_o}{c^2}\right)\mathbf{U}$$

where,

- Four-momentum $\mathbf{P} = \left(\frac{E}{c}, \vec{\mathbf{p}}\right)$.
- Four-wavevector $\mathbf{K} = \left(\frac{\omega}{c}, \vec{\mathbf{k}}\right) = \left(\frac{\omega}{c}, \frac{\omega}{v_p}\hat{\mathbf{n}}\right)$.
- Four-velocity $\mathbf{U} = \gamma(c, \vec{\mathbf{u}}) = \gamma(c, v_g\hat{\mathbf{n}})$.

Interpretations

The physical reality underlying de Broglie waves is a subject of ongoing debate. Some theories treat either the particle or the wave aspect as its fundamental nature, seeking to explain the other as an emergent property. Some, such as the hidden variable theory, treat the wave and the particle as distinct entities. Yet others propose some intermediate entity that is neither quite wave nor quite particle but only appears as such when we measure one or the other property. The Copenhagen interpretation states that the nature of the underlying reality is unknowable and beyond the bounds of scientific inquiry.

Schrödinger's quantum mechanical waves are conceptually different from ordinary physical waves such as water or sound. Ordinary physical waves are characterized by undulating real-number 'displacements' of dimensioned physical variables at each point of ordinary physical space at each instant of time. Schrödinger's "waves" are characterized by the undulating value of a dimensionless complex number at each point of an abstract multi-dimensional space, for example of configuration space.

At the Fifth Solvay Conference in 1927, Max Born and Werner Heisenberg reported as follows:

> "If one wishes to calculate the probabilities of excitation and ionization of atoms then one must introduce the coordinates of the atomic electrons as variables on an equal footing with those of the colliding electron. The waves then propagate no longer in three-dimensional space but in multi-dimensional configuration space. From this one sees that the quantum mechanical waves are indeed something quite different from the light waves of the classical theory."

At the same conference, Erwin Schrödinger reported likewise:

> "Under the name 'wave mechanics', at present two theories are being carried on, which are indeed closely related but not identical. The first, which follows on directly from the famous doctoral thesis by L. de Broglie, concerns waves in three-dimensional space. Because of the strictly relativistic treatment that is adopted in this version from the outset, we shall refer to it as the *four-dimensional* wave mechanics. The other theory is more remote from Mr de Broglie's original ideas, insofar as it is based on a wave-like process in the space of *position coordinates* (q-space) of an arbitrary mechanical system. We shall therefore call it the *multi-dimensional* wave mechanics. Of course this use of the q-space is to be seen only as a mathematical tool, as it is often applied also in the old mechanics; ultimately, in this version also, the process to be described is one in space and time. In truth, however, a complete unification of the two conceptions has not yet been achieved. Anything over and above the motion of a single electron could be treated so far only in the *multi*-dimensional version; also, this

> is the one that provides the mathematical solution to the problems posed by the Heisenberg-Born matrix mechanics."

In 1955, Heisenberg reiterated this:

> "An important step forward was made by the work of Born in the summer of 1926. In this work, the wave in configuration space was interpreted as a probability wave, in order to explain collision processes on Schrödinger's theory. This hypothesis contained two important new features in comparison with that of Bohr, Kramers and Slater. The first of these was the assertion that, in considering "probability waves", we are concerned with processes not in ordinary three-dimensional space, but in an abstract configuration space (a fact which is, unfortunately, sometimes overlooked even today); the second was the recognition that the probability wave is related to an individual process."

The "displaced quantity" of the Schrödinger wave has values that are dimensionless complex numbers. One may ask what is the physical meaning of those numbers. According to Heisenberg, rather than being of some ordinary physical quantity such as, for example, Maxwell's electric field intensity, or mass density, the Schrödinger-wave packet's "displaced quantity" is probability amplitude. He wrote that instead of using the term 'wave packet', it is preferable to speak of a probability packet. The probability amplitude supports calculation of probability of location or momentum of discrete particles. Heisenberg recites Duane's account of particle diffraction by probabilistic quantal translation momentum transfer, which allows, for example in Young's two-slit experiment, each diffracted particle probabilistically to pass discretely through a particular slit. Thus one does not need necessarily think of the matter wave, as it were, as 'composed of smeared matter'.

These ideas may be expressed in ordinary language as follows. In the account of ordinary physical waves, a 'point' refers to a position in ordinary physical space at an instant of time, at which there is specified a 'displacement' of some physical quantity. But in the account of quantum mechanics, a 'point' refers to a configuration of the system at an instant of time, every particle of the system being in a sense present in every 'point' of configuration space, each particle at such a 'point' being located possibly at a different position in ordinary physical space. There is no explicit definite indication that, at an instant, this particle is 'here' and that particle is 'there' in some separate 'location' in configuration space. This conceptual difference entails that, in contrast to de Broglie's pre-quantum mechanical wave description, the quantum mechanical probability packet description does not directly and explicitly express the Aristotelian idea, referred to by Newton, that causal efficacy propagates through ordinary space by contact, nor the Einsteinian idea that such propagation is no faster than light. In contrast, these ideas are so expressed in the classical wave account, through the Green's function, though it is inadequate for the observed quantal phenomena. The physical reasoning for this was first recognized by Einstein.

De Broglie's Phase Wave and Periodic Phenomenon

De Broglie's thesis started from the hypothesis, "that to each portion of energy with a proper mass m_0 one may associate a periodic phenomenon of the frequency v_0, such that one finds: $hv_0 = m_0c^2$. The frequency v_0 is to be measured, of course, in the rest frame of the energy packet. This hypothesis is the basis of our theory."

De Broglie followed his initial hypothesis of a periodic phenomenon, with frequency v_0, associated with the energy packet. He used the special theory of relativity to find, in the frame of the observer of the electron energy packet that is moving with velocity v, that its frequency was apparently reduced to:

$$f = v_0\sqrt{1-\frac{v^2}{c^2}}.$$

Then,

$$\lambda f = E/p = v_p.$$

using the same notation as above. The quantity v_p is the velocity of what de Broglie called the "phase y wave". Its wavelength is λ and frequency f. De Broglie reasoned that his hypothetical intrinsic particle periodic phenomenon is in phase with that phase wave. This was his basic matter wave conception. He noted, as above, that $v_p > c$, and the phase wave does not transfer energy.

While the concept of waves being associated with matter is correct, de Broglie did not leap directly to the final understanding of quantum mechanics with no missteps. There are conceptual problems with the approach that de Broglie took in his thesis that he was not able to resolve, despite trying a number of different fundamental hypotheses in different papers published while working on, and shortly after publishing, his thesis. These difficulties were resolved by Erwin Schrödinger, who developed the wave mechanics approach, starting from a somewhat different basic hypothesis.

UNCERTAINITY PRINCIPLE

The uncertainty principle,' sometimes called the Heisenberg uncertainty principle, states that interaction and mechanical action come in quanta, that is, in discrete units. While one can detect an integral number of quanta, one cannot detect one half or two thirds of a quantum. The quantum of interaction joins the quanta of energy, time, and space in the current description of the world. The size of all these quanta is small—which is why they appear continuous—and determined by the size of Planck's Constant.

Thus, in quantum physics, the outcome of even an ideal measurement of a system is not deterministic, but instead is characterized by a probability distribution, and the larger the associated standard deviation is, the more "uncertain" one might say that characteristic is for the system. The Heisenberg uncertainty principle gives a lower bound on the product of the standard deviations of position and momentum for a system, implying that it is impossible to have a particle that has an arbitrarily well-defined position and momentum simultaneously. More precisely, the product of the standard deviations $\Delta x \Delta p \geq \hbar/2$, where $\hbar$ is Planck's constant, h, divided by 2π (the circumference of the unit circle).

The principle generalizes to many other pairs of quantities besides position and momentum. Examples of such complementary pairs are angular momentum about two different axes, and mass-energy in time, and can be derived directly from the axioms of quantum mechanics.

The uncertainties in question are characteristic of the nature of the physical world. In any real-world measurement, there will also be additional uncertainties created by the non-ideal and imperfect measurement process. The uncertainty principle holds true regardless of whether the measurements are ideal (sometimes called von Neumann measurements) or non-ideal (Landau measurements). Note also that the product of the uncertainties, of order 10^{-35} Joule-seconds, is so small that the uncertainty principle has negligible effect on objects of macroscopic scale and things appear to be continuous rather than quantal, rather like the pixels of a photo on a computer screen that, from a distance, blend into a continuum.

The uncertainty principle was an important step in the development of quantum mechanics when it was discovered by Werner Heisenberg in 1927. It is connected to the observer effect—that observation requires interaction, and interaction involves a quantum of change, an unavoidable "disturbance" of the thing being observed.

Wave-particle Duality

A fundamental postulate of quantum mechanics, which manifests itself in the Heisenberg Uncertainty Principle, is that no physical phenomenon can be (to arbitrary accuracy) described as a "classic point particle" or as a wave but rather the microphysical situation is best described in terms of wave-particle duality.

The Heisenberg uncertainty principle is a consequence of this picture. The amplitude of the wave associated with a particle corresponds to its position, and the wavelength (more precisely, its Fourier transform) is inversely proportional to momentum. In order to localize the wave so as to have a sharp peak (that is, a small position uncertainty), it is necessary to incorporate waves with very short wavelengths, corresponding to high momenta in all directions, and therefore a large momentum uncertainty. Indeed, the Heisenberg Uncertainty Principle is equivalent to a theorem in functional analysis that the standard deviation of the squared absolute value of a function, times the standard deviation of the squared absolute value of its Fourier transform, is at least $1/(16\pi^2)$.

A helpful analogy can be drawn between the wave associated with a quantum-mechanical particle and a more familiar wave, the time-varying signal associated with, say, a sound wave. The main difference in the math is that while sound waves can be described by real numbers, the waves of quantum mechanics require the use of complex numbers involving the square root of minus one. This is called an "internal" wave function to distinguish it from external, real-numbered waves.

Uncertainty Principle versus Observer Effect

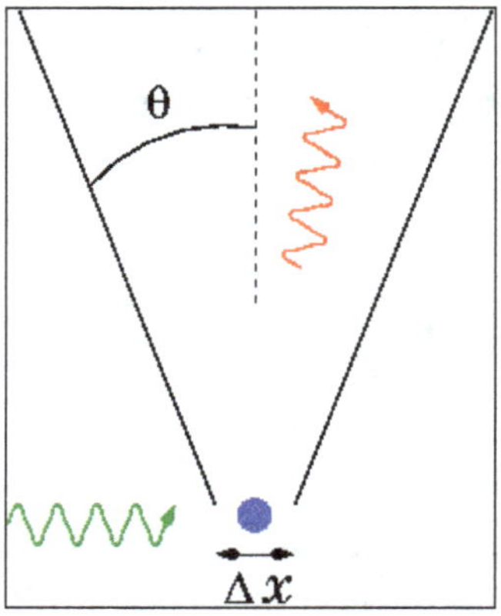

Heisenberg's gamma-ray microscope for locating an electron (shown in blue). The incoming gamma ray (shown in green) is scattered by the electron up into the microscope's aperture angle θ. The scattered gamma-ray is shown in red. Classical optics shows that the electron position can be resolved only up to an uncertainty Δx that depends on θ and the wavelength λ of the incoming light.

The uncertainty principle in quantum mechanics is connected to the observer effect—that a measurement inevitably involves a whole quantum and subsequent change. Indeed, Heisenberg himself developed his concepts starting with this view. A measurement was visualized as a physical disturbance inflicted directly on the measured system, being sometimes illustrated as a thought experiment called Heisenberg's microscope. For instance, when measuring the position of an electron, one imagines shining a light on it, thus disturbing the electron and producing the quantum mechanical uncertainties in its position.

It is understood that the uncertainties in the system exist prior to and independent of any the measurement, and the uncertainty principle is therefore more fundamental than the observer effect.

Complementary Pairs

A few of the more common pairs in the uncertainty principle are:

- There is an uncertainty relation between the position and momentum of an object:

$$\Delta x_i \Delta p_i \geq \frac{\hbar}{2}$$

- Between angular position and angular momentum of an object:

  $$\Delta O_i \Delta J_i \geq \frac{\hbar}{2}$$

- Between two orthogonal components of the total angular momentum operator of an object:

  $$\Delta J_i \Delta J_j \geq \frac{\hbar}{2}\left|\langle J_k \rangle\right|$$

  where i, j, k are distinct and J_i denotes angular momentum along the x_i axis.

- Between the number of electrons in a superconductor and the phase of its Ginzburg-Landau order parameter:

  $$\Delta N \Delta \phi \geq 1$$

Interpretations

The Uncertainty Principle was developed as an answer to the question: How does one measure the location of an electron around a nucleus?

In the summer of 1922, Heisenberg met Niels Bohr, the founding father of quantum mechanics, and in September 1924, Heisenberg went to Copenhagen, where Bohr had invited him as a research associate and later as his assistant. In 1925, Werner Heisenberg laid down the basic principles of a complete quantum mechanics. In his new matrix theory, he replaced classical commuting variables with non-commuting ones. Heisenberg's paper marked a radical departure from previous attempts to solve atomic problems by making use of observable quantities only. He wrote in a 1925 letter, "My entire meager efforts go toward killing off and suitably replacing the concept of the orbital paths that one cannot observe." Rather than struggle with the complexities of three-dimensional orbits, Heisenberg dealt with the mechanics of a one-dimensional vibrating system, an anharmonic oscillator. The result was formulae in which quantum numbers were related to observable radiation frequencies and intensities. In March 1926, working in Bohr's institute, Heisenberg formulated the principle of uncertainty, thereby laying the foundation of what became known as the Copenhagen interpretation of quantum mechanics.

Albert Einstein was not happy with the uncertainty principle, and he challenged Niels Bohr and Werner Heisenberg with a famous thought experiment: Fill a box with a radioactive material which randomly emits radiation. The box has a shutter, which is opened and soon thereafter shut by a clock at a precise time, thereby allowing some radiation to escape. So the time is already known with precision. But the experimenter still wants to measure the conjugate variable energy precisely. Einstein proposed doing this by weighing the box before and after. The equivalence between mass and energy from special relativity will allow the experimenter to determine precisely how much energy was left in the box.

Bohr countered as follows: Should energy leave, then the now lighter box will rise slightly on the scale. That changes the position of the clock. Thus, the clock deviates from a stationary reference frame, and by general relativity, its measurement of time will be different from the observer's, leading to some unavoidable margin of error. In fact, a detailed analysis shows that the imprecision is correctly given by Heisenberg's relation.

The term, "Copenhagen interpretation of quantum mechanics" was often used interchangeably with and as a synonym for Heisenberg's Uncertainty Principle by detractors who believed in fate and determinism and saw the common features of the Bohr-Heisenberg theories as a threat. Within the widely but not universally accepted Copenhagen interpretation of quantum mechanics (that is, it was not accepted by Einstein or other physicists such as Alfred Lande), the uncertainty principle is taken to mean that on an elementary level, the physical universe does not exist in a deterministic form—but rather as a collection of probabilities, or potentials. For example, the pattern (probability distribution) produced by millions of photons passing through a diffraction slit can be calculated using quantum mechanics, but the exact path of each photon cannot be predicted by any known method. The Copenhagen interpretation holds that it cannot be predicted by any method, not even with theoretically infinitely precise measurements.

It is this interpretation that Einstein was questioning when he said, "I cannot believe that God would choose to play dice with the universe." Bohr, who was one of the authors of the Copenhagen interpretation, responded, "Einstein, don't tell God what to do." Niels Bohr himself acknowledged that quantum mechanics and the uncertainty principle were counter-intuitive when he stated, "Anyone who is not shocked by quantum theory has not understood a single word."

The basic debate between Einstein and Bohr (including Heisenberg's Uncertainty Principle) was that Einstein was in essence saying: "Of course, we can know where something is; we can know the position of a moving particle if we know every possible detail, and therefore by extension, we can predict where it will go." Bohr and Heisenberg were saying: "We can only know the probable position of a moving particle, therefore by extension, we can only know its probable destination; we can never know with absolute certainty where it will go."

Einstein was convinced that this interpretation was in error. His reasoning was that all previously known probability distributions arose from deterministic events. The distribution of a flipped coin or a rolled die can be described with a probability distribution (50 percent heads, 50 percent tails), but this does not mean that their physical motions are unpredictable. Ordinary mechanics can be used to calculate exactly how each coin will land, if the forces acting on it are known. And the heads/tails distribution will still line up with the probability distribution (given random initial forces).

Einstein assumed that there are similar hidden variables in quantum mechanics which underlie the observed probabilities and that these variables, if known, would show that there was what Einstein termed "local realism," a description opposite to the uncertainty principle, being that all objects must already have their properties before they are observed or measured. For the greater part of the twentieth century, there were many such hidden variable theories proposed, but in 1964 John Bell theorized the Bell inequality to counter them, which postulated that although the behavior of an individual particle is random, it is also correlated with the behavior of other particles. Therefore, if the uncertainty principle is the result of some deterministic process in which a particle has local realism, it must be the case that particles at great distances instantly transmit information to each other to ensure that the correlations in behavior between particles occur. The interpretation of Bell's theorem explicitly prevents any local hidden variable theory from holding true because it shows the necessity of a system to describe correlations between objects. The implication is, if a hidden local variable is the cause of particle 1 being at a position, then a second hidden local variable would be responsible for particle 2 being in its own position—and there is no system to correlate the behavior between them. Experiments have demonstrated that there is a correlation. In the years following, Bell's theorem was tested and has held up experimentally time and time again, and these experiments are in a sense the clearest experimental confirmation of quantum mechanics. It is worth noting that Bell's theorem only applies to local hidden variable theories; non-local hidden variable theories can still exist (which some, including Bell, think is what can bridge the conceptual gap between quantum mechanics and the observable world).

Whether Einstein's view or Heisenberg's view is true or false is not a directly empirical matter. One criterion by which scientists may judge the success of a scientific theory is the explanatory power it provides, and to date, it seems that Heisenberg's view has been the better at explaining physical subatomic phenomena.

Mathematics

Applying the ideas of special relativity to the position-momentum uncertainty principle, many physicists, like Niels Bohr, postulated that the following relation ought to exist:

$$\Delta E \Delta t \geq \frac{\hbar}{2},$$

But it was not immediately obvious how Δt should be defined (since time is not treated as an operator). In 1926, Dirac offered a precise definition and derivation of this uncertainty relation, as coming from a relativistic quantum theory of "events." But the better-known, more widely-used, correct formulation was given only in 1945, by L. I. Mandelshtam and

I. E. Tamm, as follows. For a quantum system in a non-stationary state $|\psi\rangle$ and an observable B represented by a self-adjoint operator $\hat{B}$, the following formula holds:

$$\Delta_\psi E \frac{\Delta_\psi B}{\left|\frac{\mathrm{d}\langle \hat{B}\rangle}{\mathrm{d}t}\right|} \geq \frac{\hbar}{2}$$

where $\Delta_\psi E$ is the standard deviation of the energy operator in the state $|\psi\rangle$, $\Delta_\psi B$ stands for the standard deviation of the operator $\hat{B}$ and $\langle \hat{B}\rangle$ is the expectation value of $\hat{B}$ in that state. Although, the second factor in the left-hand side has dimension of time, it is different from the time parameter that enters Schrödinger equation. It is a lifetime of the state $|\psi\rangle$ with respect to the observable B. In other words, this is the time after which the expectation value $\langle \hat{B}\rangle$ changes appreciably.

The energy-time uncertainty principle has important implications for spectroscopy. Since excited states have a finite lifetime, they do not all release exactly the same energy when they decay; rather, the spectroscopic peaks are actually bumps with a finite width (called natural linewidth), with the center of the bump corresponding to the true energy of the excited state. For fast-decaying states, the linewidth makes it difficult to accurately measure this true energy, and indeed, researchers have even used microwave cavities to slow down the decay-rate, in order to get sharper peaks and more accurate energy measurements.

One particularly famous false formulation of the energy-time uncertainty principle says that the energy of a quantum system measured over the time interval Δt has to be inaccurate, with the inaccuracy ΔE given by the inequality $\Delta E \Delta t \geq \hbar/2$. This formulation was explicitly invalidated by Y. Aharonov and D. Bohm in 1961. Indeed, one can actually determine the accurate energy of a quantum system in an arbitrarily short interval of time. Moreover, as recent research indicates, for quantum systems with discrete energy spectra the product $\Delta E \Delta t$ is bounded from above by a statistical noise that in fact vanishes if sufficiently many identical copies of the system are used. This vanishing upper bound certainly removes the possibility of a lower bound, again disproving this false formulation of the energy-time uncertainty principle.

Generalization, Precise Formulation and Robertson-Schrödinger Relation

Measurements of position and momentum taken in several identical copies of a system in a given state will each vary according to a probability distribution characteristic of the system's state. This is the fundamental postulate of quantum mechanics.

If we compute the standard deviations Δx and Δp of the position and momentum measurements, then:

$$\Delta x \Delta p \geq \frac{\hbar}{2}$$

where,

$\hbar$ is the reduced Planck's constant (Planck's constant divided by 2π).

More generally, given any Hermitian operators A and B, and a system in the state ψ, there are probability distributions associated with the measurement of each of A and B, giving rise to standard deviations $\Delta_\psi A$ and $\Delta_\psi B$. Then:

$$\Delta_\psi A \Delta_\psi B \geq \frac{1}{2}\left|\langle[A,B]\rangle_\psi\right|$$

where the operator [A,B] = AB - BA denotes the commutator of A and B, and $\langle X\rangle_\psi$ denotes expectation value. This inequality is called the Robertson-Schrödinger relation, and includes the Heisenberg Uncertainty Principle as a special case. It was first pointed out in 1930, by Howard Percy Robertson and (independently) by Erwin Schrödinger.

Derivation

The uncertainty principle has a straightforward mathematical derivation. The key step is an application of the Cauchy-Schwarz inequality, one of the most useful theorems of linear algebra.

For two arbitrary Hermitian operators A: H → H and B: H → H, and any element x of H, then:

$$\langle BAx \mid x\rangle = \langle Ax \mid Bx\rangle = \langle Bx \mid Ax\rangle^*$$

In an inner product space the Cauchy-Schwarz inequality holds:

$$\left|\langle Bx \mid Ax\rangle\right|^2 \leq \| Ax \|^2 \| Bx \|^2$$

Rearranging this formula leads to:

$$\begin{aligned}
\| Ax \|^2 \| Bx \|^2 \geq \left|\langle Bx \mid Ax\rangle\right|^2 &\geq \left|\mathrm{Im}\{\langle Bx \mid Ax\rangle\}\right|^2 \\
&= \frac{1}{4}\left|2\,\mathrm{Im}\{\langle Bx \mid Ax\rangle\}\right|^2 \\
&= \frac{1}{4}\left|\langle Bx \mid Ax\rangle - \langle Bx \mid Ax\rangle^*\right|^2 \\
&= \frac{1}{4}\left|\langle Bx \mid Ax\rangle - \langle Ax \mid Bx\rangle\right|^2 \\
&= \frac{1}{4}\left|\langle ABx \mid x\rangle - \langle BAx \mid x\rangle\right|^2 \\
&= \frac{1}{4}\left|\langle (AB - BA)x \mid x\rangle\right|^2
\end{aligned}$$

This gives one form of the Robertson-Schrödinger relation:

$$\frac{1}{4}|\langle [A,B]x \mid x\rangle|^2 \leq \|Ax\|^2 \|Bx\|^2,$$

where the operator [A,B] = AB - BA denotes the commutator of A and B.

To make the physical meaning of this inequality more directly apparent, it is often written in the equivalent form:

$$\Delta_\psi A \Delta_\psi B \geq \frac{1}{2}\left|\left\langle [A,B]\right\rangle_\psi\right|$$

where,

$$\left\langle X\right\rangle_\psi = \left\langle \psi \mid X\psi\right\rangle$$

is the operator mean of observable X in the system state ψ and,

$$\Delta_\psi X = \sqrt{\langle X^2\rangle_\psi - \langle X\rangle_\psi^2}$$

is the operator standard deviation of observable X in the system state ψ. This formulation can be derived from the above formulation by plugging in $A - \langle A\rangle_\psi$ for A and $B - \langle B\rangle_\psi$ for B, and using the fact that:

$$[A,B] = [A - \langle A\rangle, B - \langle B\rangle]$$

This formulation acquires its physical interpretation, indicated by the suggestive terminology "mean" and "standard deviation," due to the properties of measurement in quantum mechanics. Particular uncertainty relations, such as position-momentum, can usually be derived by a straightforward application of this inequality.

References

- Martins, Roberto de Andrade (2010). "Louis de Broglie's Struggle with the Wave-Particle Dualism, 1923-1925". Quantum History Project, Fritz Haber Institute of the Max Planck Society and the Max Planck Institute for the History of Science. Retrieved 3 January 2015
- Quantum-mechanics-physics, science: britannica.com, Retrieved 7 April, 2020
- J. P. Mcevoy; Oscar Zarate (2004). Introducing Quantum Theory. Totem Books. Pp. 110–114. ISBN 978-1-84046-577-8
- Uncertainty-principle, entry: newworldencyclopedia.org, Retrieved 5 February, 2020
- Z.Y.Wang (2016). "Generalized momentum equation of quantum mechanics". Optical and Quantum Electronics. 48 (2). doi:10.1007/s11082-015-0261-8

2

The Wave Function

The wave function is defined as the variable quantity which describes the wave characteristics of a particle mathematically. Vector space structure, inner product, Hilbert space, Schrödinger equation, etc. are some of its concepts. The topics elaborated in this chapter will help in gaining a better perspective of wave function.

Wave function, in quantum mechanics is a variable quantity that mathematically describes the wave characteristics of a particle. The value of the wave function of a particle at a given point of space and time is related to the likelihood of the particle's being there at the time. By analogywith waves such as those of sound, a wave function, designated by the letter psi, Ψ, may be thought of as an expression for the amplitude of the particle wave (or de Broglie wave), although for such waves amplitude has no physical significance. The square of the wave function, Ψ^2, however, does have physical significance: the probability of finding the particle described by a specific wave function Ψ at a given point and time is proportional to the value of Ψ^2.

Born's Interpretation

The wave function ψ itself has no physical significance but the square of its absolute magnitude $|\psi^2|$ has significance when evaluated at a particular point and at a particular time $|\psi^2|$ gives the probability of finding the particle there at that time.

The wave function ψ(x,t) is a quantity such that the product:

$$P(x,t) = \psi^*(x,t)\psi(x,t)$$

Is the probability per unit length of finding the particle at the position x at time t. P(x,t) is the probability density and $\psi^*(x,t)$ is complex conjugate of ψ(x,t). Hence the probability of finding the particle is large wherever ψ is large and vice-versa.

Normalization Condition

The probability per unit length of finding the particle at position x at time t is:

$$P = \psi^*(x,t)\psi(x,t)$$

So, probability of finding the particle in the length dx is:

$$\mathrm{Pdx} = \psi^*(x,t)\psi(x,t)\,dx$$

Total probability of finding the particle somewhere along x-axis is:

$$\int \mathrm{pdx} = \int \psi^*(x,t)\psi(x,t)\,dx$$

If the particle exists, it must be somewhere on the x-axis . so the total probability of finding the particle must be unity i.e.:

$$\int \psi^*(x,t)\psi(x,t)\,dx = 1$$

This is called the normalization condition. So a wave function ψ(x,t) is said to be normalized if it satisfies the condition in the above equation.

Orthogonal Wave Functions

Consider two different wave functions ψ_m and ψ_n such that both satisfies Schrodinger equation. These two wave functions are said to be orthogonal if they satisfy the conditions.

Or,

$$\int \psi_n^*(x,t)\ \psi_m(x,t)\ dV = 0 \text{ for } n \neq m]$$

$$\int \psi_n^*(x,t)\ \psi_m(x,t)\ dV = 0 \text{ for } m \neq n\ ]$$

If both the wave functions are simultaneously normal then:

$$\int \psi_m \psi_m^*\, dV = 1 = \int \psi_n \psi_n^*\, dV$$

Orthonormal Wave Functions

The sets of wave functions, which are both normalized as well as orthogonal are called orthonormal wave functions.

$$\int \psi_m^* \psi_n dV = \{ \text{ o if } \ m \neq n$$

$$= [1 \text{ if } m = n\,.$$

One Spinless Particle in One Dimension

The real parts of position wave function Ψ(*x*) and momentum wave function Φ(*p*), and corresponding probability densities |Ψ(*x*)|² and |Φ(*p*)|², for one spin-0 particle in one *x* or *p* dimension. The colour opacity of the particles corresponds to the probability density (*not* the wave function) of finding the particle at position *x* or momentum *p*.

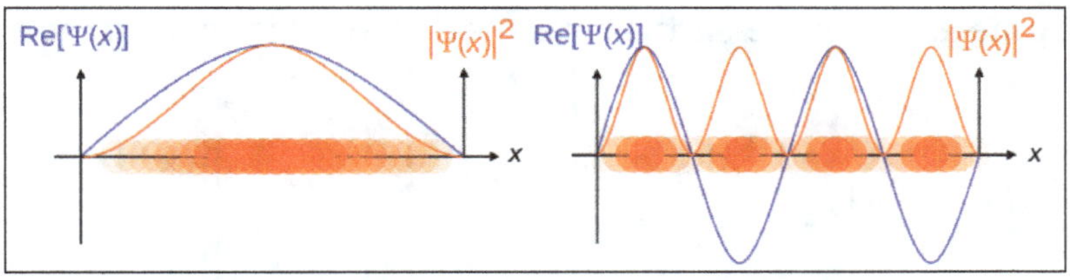

Standing waves for a particle in a box, examples of stationary states.

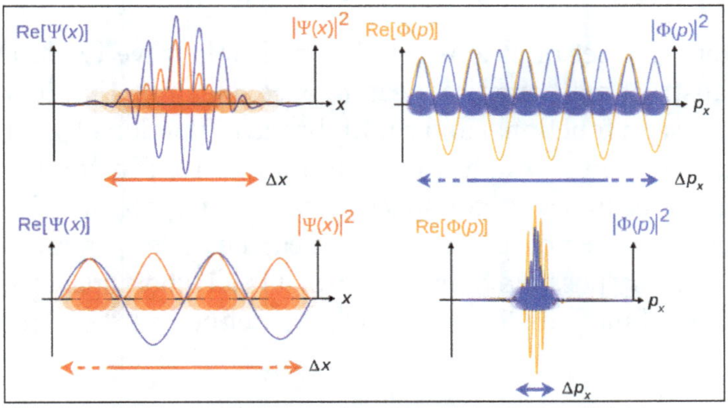

Travelling waves of a free particle.

For now, consider the simple case of a non-relativistic single particle, without spin, in one spatial dimension.

Position-space Wave Functions

The state of such a particle is completely described by its wave function:

$$\Psi(x,t),$$

where x is position and t is time. This is a complex-valued function of two real variables x and t.

For one spinless particle in 1d, if the wave function is interpreted as a probability amplitude, the square modulus of the wave function, the positive real number:

$$|\Psi(x,t)|^2 = \Psi(x,t)^* \Psi(x,t) = \rho(x,t),$$

is interpreted as the probability density that the particle is at x. The asterisk indicates the complex conjugate. If the particle's position is measured, its location cannot be determined from the wave function, but is described by a probability distribution. The probability that its position x will be in the interval $a \leq x \leq b$ is the integral of the density over this interval:

$$P_{a\leq x\leq b}(t) = \int_a^b |\Psi(x,t)|^2 \, dx$$

where t is the time at which the particle was measured. This leads to the normalization condition:

$$\int_{-\infty}^{\infty} |\Psi(x,t)|^2 \, dx = 1,$$

because if the particle is measured, there is 100% probability that it will be *somewhere*.

For a given system, the set of all possible normalizable wave functions (at any given time) forms an abstract mathematical vector space, meaning that it is possible to add together different wave functions, and multiply wave functions by complex numbers. Technically, because of the normalization condition, wave functions form a projective space rather than an ordinary vector space. This vector space is infinite-dimensional, because there is no finite set of functions which can be added together in various combinations to create every possible function. Also, it is a Hilbert space, because the inner product of two wave functions Ψ_1 and Ψ_2 can be defined as the complex number (at time *t*):

$$(\Psi_1, \Psi_2) = \int_{-\infty}^{\infty} \Psi_1^*(x,t)\Psi_2(x,t) \, dx,$$

Although the inner product of two wave functions is a complex number, the inner product of a wave function Ψ with itself:

$$(\Psi, \Psi) = \|\Psi\|^2,$$

is *always* a positive real number. The number $||\Psi||$ (not $||\Psi||^2$) is called the norm of the wave function Ψ.

If $(\Psi, \Psi) = 1$, then Ψ is normalized. If Ψ is not normalized, then dividing by its norm gives the normalized function $\Psi/||\Psi||$. Two wave functions Ψ_1 and Ψ_2 are orthogonal if $(\Psi_1, \Psi_2) = 0$. If they are normalized *and* orthogonal, they are orthonormal. Orthogonality (hence also orthonormality) of wave functions is not a necessary condition wave functions must satisfy, but is instructive to consider since this guarantees linear independence of the functions. In a linear combination of orthogonal wave functions Ψ_n we have:

$$\Psi = \sum_n a_n \Psi_n, \quad a_n = \frac{(\Psi_n, \Psi)}{(\Psi_n, \Psi_n)},$$

If the wave functions Ψ_n were nonorthogonal, the coefficients would be less simple to obtain.

In the Copenhagen interpretation, the modulus squared of the inner product (a complex number) gives a real number:

$$|(\Psi_1, \Psi_2)|^2 = P(\Psi_2 \to \Psi_1),$$

which, assuming both wave functions are normalized, is interpreted as the probability of the wave function Ψ_2 "collapsing" to the new wave function Ψ_1 upon measurement of an observable, whose eigenvalues are the possible results of the measurement, with Ψ_1 being an eigenvector of the resulting eigenvalue. This is the Born rule, and is one of the fundamental postulates of quantum mechanics.

At a particular instant of time, all values of the wave function $\Psi(x, t)$ are components of a vector. There are uncountably infinitely many of them and integration is used in place of summation. In Bra–ket notation, this vector is written:

$$|\Psi(t)\rangle = \int \Psi(x,t)\,|x\rangle\, dx$$

and is referred to as a "quantum state vector", or simply "quantum state". There are several advantages to understanding wave functions as representing elements of an abstract vector space:

- All the powerful tools of linear algebra can be used to manipulate and understand wave functions. For example:
  - Linear algebra explains how a vector space can be given a basis, and then any vector in the vector space can be expressed in this basis. This explains the relationship between a wave function in position space and a wave function in momentum space and suggests that there are other possibilities too.
  - Bra–ket notation can be used to manipulate wave functions.
- The idea that quantum states are vectors in an abstract vector space is completely general in all aspects of quantum mechanics and quantum field theory, whereas the idea that quantum states are complex-valued "wave" functions of space is only true in certain situations.

The time parameter is often suppressed, and will be in the following. The x coordinate is a continuous index. The $|x\rangle$ are the basis vectors, which are orthonormal so their inner product is a delta function;

$$\langle x' | x\rangle = \delta(x' - x)$$

Thus,

$$\langle x' | \Psi\rangle = \int \Psi(x)\langle x' | x\rangle\, dx = \Psi(x')$$

and,

$$|\Psi\rangle = \int |x\rangle\langle x | \Psi\rangle\, dx = \left(\int |x\rangle\langle x|\, dx\right)|\Psi\rangle$$

which illuminates the identity operator,

$$I = \int |x\rangle\langle x|\, dx.$$

Finding the identity operator in a basis allows the abstract state to be expressed explicitly in a basis, and more (the inner product between two state vectors, and other operators for observables, can be expressed in the basis).

Momentum-space Wave Functions

The particle also has a wave function in momentum space:

$$\Phi(p,t)$$

where p is the momentum in one dimension, which can be any value from $-\infty$ to $+\infty$, and t is time.

Analogous to the position case, the inner product of two wave functions $\Phi_1(p, t)$ and $\Phi_2(p, t)$ can be defined as:

$$(\Phi_1,\Phi_2)=\int_{-\infty}^{\infty}\Phi_1^*(p,t)\Phi_2(p,t)\,dp.$$

One particular solution to the time-independent Schrödinger equation is:

$$\Psi_p(x)=e^{ipx/\hbar},$$

a plane wave, which can be used in the description of a particle with momentum exactly p, since it is an eigenfunction of the momentum operator. These functions are not normalizable to unity (they aren't square-integrable), so they are not really elements of physical Hilbert space. The set:

$$\{\Psi_p(x,t),-\infty\leq p\leq\infty\}$$

forms what is called the momentum basis. This "basis" is not a basis in the usual mathematical sense. For one thing, since the functions aren't normalizable, they are instead normalized to a delta function,

$$(\Psi_p,\Psi_{p'})=\delta(p-p').$$

For another thing, though they are linearly independent, there are too many of them (they form an uncountable set) for a basis for physical Hilbert space. They can still be used to express all functions in it using Fourier transforms.

Relations between Position and Momentum Representations

The x and p representations are:

$$|\Psi\rangle=I|\Psi\rangle=\int|x\rangle\langle x|\Psi\rangle dx=\int\Psi(x)|x\rangle dx,$$

$$|\Psi\rangle=I|\Psi\rangle=\int|p\rangle\langle p|\Psi\rangle dp=\int\Phi(p)|p\rangle dp.$$

Now take the projection of the state Ψ onto eigenfunctions of momentum using the last expression in the two equations:

$$\int \Psi(x)\langle p \mid x\rangle dx = \int \Phi(p')\langle p \mid p'\rangle dp' = \int \Phi(p')\delta(p-p')dp' = \Phi(p).$$

Then utilizing the known expression for suitably normalized eigenstates of momentum in the position representation solutions of the free Schrödinger equation:

$$\langle x \mid p\rangle = p(x) = \frac{1}{\sqrt{2\pi\hbar}} e^{\frac{i}{\hbar}px} \Rightarrow \langle p \mid x\rangle = \frac{1}{\sqrt{2\pi\hbar}} e^{-\frac{i}{\hbar}px},$$

one obtains:

$$\Phi(p) = \frac{1}{\sqrt{2\pi\hbar}} \int \Psi(x) e^{-\frac{i}{\hbar}px} dx.$$

Likewise, using eigenfunctions of position:

$$\Psi(x) = \frac{1}{\sqrt{2\pi\hbar}} \int \Phi(p) e^{\frac{i}{\hbar}px} dp.$$

The position-space and momentum-space wave functions are thus found to be Fourier transforms of each other. The two wave functions contain the same information, and either one alone is sufficient to calculate any property of the particle. As representatives of elements of abstract physical Hilbert space, whose elements are the possible states of the system under consideration, they represent the same state vector, hence *identical physical states*, but they are not generally equal when viewed as square-integrable functions.

In practice, the position-space wave function is used much more often than the momentum-space wave function. The potential entering the relevant equation (Schrödinger, Dirac, etc.) determines in which basis the description is easiest. For the harmonic oscillator, x and p enter symmetrically, so there it doesn't matter which description one uses. The same equation (modulo constants) results. From this follows, with a little bit of afterthought, a factoid: The solutions to the wave equation of the harmonic oscillator are eigenfunctions of the Fourier transform in L^2.

Other Cases

Following are the general forms of the wave function for systems in higher dimensions and more particles, as well as including other degrees of freedom than position coordinates or momentum components.

One-particle States in 3d Position Space

The position-space wave function of a single particle without spin in three spatial dimensions is similar to the case of one spatial dimension:

$$\Psi(\mathbf{r},t)$$

where $\mathbf{r}$ is the position vector in three-dimensional space, and t is time. As always $\Psi(\mathbf{r}, t)$ is a complex-valued function of real variables. As a single vector in Dirac notation,

$$|\Psi(t)\rangle = \int d^3\mathbf{r}\,\Psi(\mathbf{r},t)|\mathbf{r}\rangle$$

All the previous remarks on inner products, momentum space wave functions, Fourier transforms, and so on extend to higher dimensions.

For a particle with spin, ignoring the position degrees of freedom, the wave function is a function of spin only (time is a parameter):

$$\xi(s_z, t)$$

where s_z is the spin projection quantum number along the z axis. (The z axis is an arbitrary choice; other axes can be used instead if the wave function is transformed appropriately.) The s_z parameter, unlike $\mathbf{r}$ and t, is a *discrete variable*. For example, for a spin-1/2 particle, s_z can only be +1/2 or −1/2, and not any other value. (In general, for spin s, s_z can be $s, s-1, \ldots, -s+1, -s$). Inserting each quantum number gives a complex valued function of space and time, there are $2s+1$ of them. These can be arranged into a column vector:

$$\xi = \begin{bmatrix} \xi(s,t) \\ \xi(s-1,t) \\ \vdots \\ \xi(-(s-1),t) \\ \xi(-s,t) \end{bmatrix} = \xi(s,t)\begin{bmatrix} 1 \\ 0 \\ \vdots \\ 0 \\ 0 \end{bmatrix} + \xi(s-1,t)\begin{bmatrix} 0 \\ 1 \\ \vdots \\ 0 \\ 0 \end{bmatrix} + \cdots + \xi(-(s-1),t)\begin{bmatrix} 0 \\ 0 \\ \vdots \\ 1 \\ 0 \end{bmatrix} + \xi(-s,t)\begin{bmatrix} 0 \\ 0 \\ \vdots \\ 0 \\ 1 \end{bmatrix}$$

In bra–ket notation, these easily arrange into the components of a vector:

$$|\xi(t)\rangle = \sum_{s_z=-s}^{s} \xi(s_z,t)|s_z\rangle$$

The entire vector ξ is a solution of the Schrödinger equation (with a suitable Hamiltonian), which unfolds to a coupled system of $2s+1$ ordinary differential equations with solutions $\xi(s, t)$, $\xi(s-1, t)$, ..., $\xi(-s, t)$. The term "spin function" instead of "wave function" is used by some authors. This contrasts the solutions to position space wave functions, the position coordinates being continuous degrees of freedom, because then the Schrödinger equation does take the form of a wave equation.

More generally, for a particle in 3d with any spin, the wave function can be written in "position–spin space" as:

$$\Psi(\mathbf{r}, s_z, t)$$

Now take the projection of the state Ψ onto eigenfunctions of momentum using the last expression in the two equations:

$$\int \Psi(x)\langle p \mid x\rangle dx = \int \Phi(p')\langle p \mid p'\rangle dp' = \int \Phi(p')\delta(p-p')dp' = \Phi(p).$$

Then utilizing the known expression for suitably normalized eigenstates of momentum in the position representation solutions of the free Schrödinger equation:

$$\langle x \mid p\rangle = p(x) = \frac{1}{\sqrt{2\pi\hbar}} e^{\frac{i}{\hbar}px} \Rightarrow \langle p \mid x\rangle = \frac{1}{\sqrt{2\pi\hbar}} e^{-\frac{i}{\hbar}px},$$

one obtains:

$$\Phi(p) = \frac{1}{\sqrt{2\pi\hbar}} \int \Psi(x) e^{-\frac{i}{\hbar}px} dx.$$

Likewise, using eigenfunctions of position:

$$\Psi(x) = \frac{1}{\sqrt{2\pi\hbar}} \int \Phi(p) e^{\frac{i}{\hbar}px} dp.$$

The position-space and momentum-space wave functions are thus found to be Fourier transforms of each other. The two wave functions contain the same information, and either one alone is sufficient to calculate any property of the particle. As representatives of elements of abstract physical Hilbert space, whose elements are the possible states of the system under consideration, they represent the same state vector, hence *identical physical states*, but they are not generally equal when viewed as square-integrable functions.

In practice, the position-space wave function is used much more often than the momentum-space wave function. The potential entering the relevant equation (Schrödinger, Dirac, etc.) determines in which basis the description is easiest. For the harmonic oscillator, x and p enter symmetrically, so there it doesn't matter which description one uses. The same equation (modulo constants) results. From this follows, with a little bit of afterthought, a factoid: The solutions to the wave equation of the harmonic oscillator are eigenfunctions of the Fourier transform in L^2.

Other Cases

Following are the general forms of the wave function for systems in higher dimensions and more particles, as well as including other degrees of freedom than position coordinates or momentum components.

One-particle States in 3d Position Space

The position-space wave function of a single particle without spin in three spatial dimensions is similar to the case of one spatial dimension:

$$\Psi(\mathbf{r},t)$$

where $\mathbf{r}$ is the position vector in three-dimensional space, and t is time. As always $\Psi(\mathbf{r},t)$ is a complex-valued function of real variables. As a single vector in Dirac notation,

$$|\Psi(t)\rangle = \int d^3\mathbf{r}\,\Psi(\mathbf{r},t)|\mathbf{r}\rangle$$

All the previous remarks on inner products, momentum space wave functions, Fourier transforms, and so on extend to higher dimensions.

For a particle with spin, ignoring the position degrees of freedom, the wave function is a function of spin only (time is a parameter):

$$\xi(s_z,t)$$

where s_z is the spin projection quantum number along the z axis. (The z axis is an arbitrary choice; other axes can be used instead if the wave function is transformed appropriately.) The s_z parameter, unlike $\mathbf{r}$ and t, is a *discrete variable*. For example, for a spin-1/2 particle, s_z can only be +1/2 or −1/2, and not any other value. (In general, for spin s, s_z can be s, $s - 1$, ... , $-s + 1$, $-s$). Inserting each quantum number gives a complex valued function of space and time, there are $2s + 1$ of them. These can be arranged into a column vector:

$$\xi = \begin{bmatrix} \xi(s,t) \\ \xi(s-1,t) \\ \vdots \\ \xi(-(s-1),t) \\ \xi(-s,t) \end{bmatrix} = \xi(s,t)\begin{bmatrix} 1 \\ 0 \\ \vdots \\ 0 \\ 0 \end{bmatrix} + \xi(s-1,t)\begin{bmatrix} 0 \\ 1 \\ \vdots \\ 0 \\ 0 \end{bmatrix} + \cdots + \xi(-(s-1),t)\begin{bmatrix} 0 \\ 0 \\ \vdots \\ 1 \\ 0 \end{bmatrix} + \xi(-s,t)\begin{bmatrix} 0 \\ 0 \\ \vdots \\ 0 \\ 1 \end{bmatrix}$$

In bra–ket notation, these easily arrange into the components of a vector:

$$|\xi(t)\rangle = \sum_{s_z=-s}^{s} \xi(s_z,t)|s_z\rangle$$

The entire vector ξ is a solution of the Schrödinger equation (with a suitable Hamiltonian), which unfolds to a coupled system of $2s + 1$ ordinary differential equations with solutions $\xi(s, t)$, $\xi(s - 1, t)$, ..., $\xi(-s, t)$. The term "spin function" instead of "wave function" is used by some authors. This contrasts the solutions to position space wave functions, the position coordinates being continuous degrees of freedom, because then the Schrödinger equation does take the form of a wave equation.

More generally, for a particle in 3d with any spin, the wave function can be written in "position–spin space" as:

$$\Psi(\mathbf{r},s_z,t)$$

and these can also be arranged into a column vector:

$$\Psi(\mathbf{r},t)=\begin{bmatrix}\Psi(\mathbf{r},s,t)\\ \Psi(\mathbf{r},s-1,t)\\ \vdots \\ \Psi(\mathbf{r},-(s-1),t)\\ \Psi(\mathbf{r},-s,t)\end{bmatrix}$$

in which the spin dependence is placed in indexing the entries, and the wave function is a complex vector-valued function of space and time only.

All values of the wave function, not only for discrete but continuous variables also, collect into a single vector:

$$|\Psi(t)\rangle=\sum_{s_z}\int d^3\mathbf{r}\,\Psi(\mathbf{r},s_z,t)|\mathbf{r},s_z\rangle$$

For a single particle, the tensor product ⊗ of its position state vector $|\psi\rangle$ and spin state vector $|\xi\rangle$ gives the composite position-spin state vector:

$$|\psi(t)\rangle\otimes|\xi(t)\rangle=\sum_{s_z}\int d^3\mathbf{r}\,\psi(\mathbf{r},t)\xi(s_z,t)|\mathbf{r}\rangle\otimes|s_z\rangle$$

with the identifications:

$$|\Psi(t)\rangle=|\psi(t)\rangle\otimes|\xi(t)\rangle$$
$$\Psi(\mathbf{r},s_z,t)=\psi(\mathbf{r},t)\xi(s_z,t)$$
$$|\mathbf{r},s_z\rangle=|\mathbf{r}\rangle\otimes|s_z\rangle$$

The tensor product factorization is only possible if the orbital and spin angular momenta of the particle are separable in the Hamiltonian operator underlying the system's dynamics (in other words, the Hamiltonian can be split into the sum of orbital and spin terms). The time dependence can be placed in either factor, and time evolution of each can be studied separately. The factorization is not possible for those interactions where an external field or any space-dependent quantity couples to the spin; examples include a particle in a magnetic field, and spin–orbit coupling.

The preceding discussion is not limited to spin as a discrete variable, the total angular momentum J may also be used. Other discrete degrees of freedom, like isospin, can expressed similarly to the case of spin above.

Many-particle States in 3d Position Space

If there are many particles, in general there is only one wave function, not a separate wave function for each particle. The fact that *one* wave function describes *many* particles is what makes quantum entanglement and the EPR paradox possible. The position-space wave function for N particles is written:

$$\Psi(\mathbf{r}_1, \mathbf{r}_2 \cdots \mathbf{r}_N, t)$$

where $\mathbf{r}_i$ is the position of the ith particle in three-dimensional space, and t is time. Altogether, this is a complex-valued function of $3N + 1$ real variables.

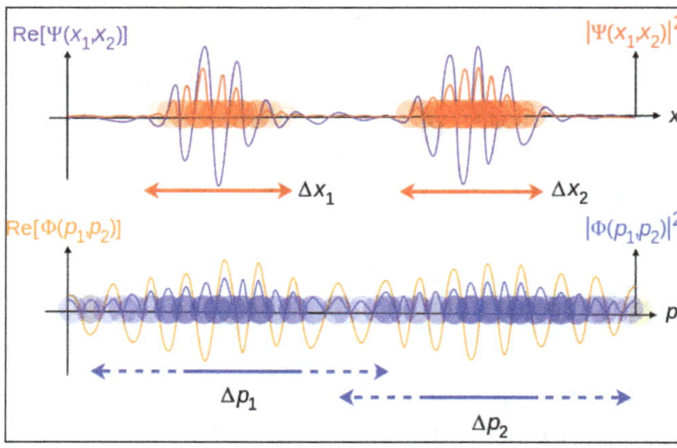

Traveling waves of two free particles, with two of three dimensions suppressed. Top is position-space wave function, bottom is momentum-space wave function, with corresponding probability densities.

In quantum mechanics there is a fundamental distinction between *identical particles* and *distinguishable* particles. For example, any two electrons are identical and fundamentally indistinguishable from each other; the laws of physics make it impossible to "stamp an identification number" on a certain electron to keep track of it. This translates to a requirement on the wave function for a system of identical particles:

$$\Psi(\ldots \mathbf{r}_a, \ldots, \mathbf{r}_b, \ldots) = \pm\Psi(\ldots \mathbf{r}_b, \ldots, \mathbf{r}_a, \ldots)$$

where the + sign occurs if the particles are *all bosons* and – sign if they are *all fermions*. In other words, the wave function is either totally symmetric in the positions of bosons, or totally antisymmetric in the positions of fermions. The physical interchange of particles corresponds to mathematically switching arguments in the wave function. The antisymmetry feature of fermionic wave functions leads to the Pauli principle. Generally, bosonic and fermionic symmetry requirements are the manifestation of particle statistics and are present in other quantum state formalisms.

For N *distinguishable* particles (no two being identical, i.e. no two having the same set of quantum numbers), there is no requirement for the wave function to be either symmetric or antisymmetric.

For a collection of particles, some identical with coordinates $\mathbf{r}_1$, $\mathbf{r}_2$, ... and others distinguishable $\mathbf{x}_1$, $\mathbf{x}_2$, ... (not identical with each other, and not identical to the aforementioned identical particles), the wave function is symmetric or antisymmetric in the identical particle coordinates $\mathbf{r}_i$ only:

$$\Psi\left(\ldots\mathbf{r}_a,\ldots,\mathbf{r}_b,\ldots,\mathbf{x}_1,\mathbf{x}_2,\ldots\right)=\pm\Psi\left(\ldots\mathbf{r}_b,\ldots,\mathbf{r}_a,\ldots,\mathbf{x}_1,\mathbf{x}_2,\ldots\right)$$

Again, there is no symmetry requirement for the distinguishable particle coordinates $\mathbf{x}_i$:

The wave function for N particles each with spin is the complex-valued function,

$$\Psi(\mathbf{r}_1,\mathbf{r}_2\cdots\mathbf{r}_N,s_{z1},s_{z2}\cdots s_{zN},t)$$

Accumulating all these components into a single vector,

$$\underbrace{|\Psi\rangle}_{\text{state vector (ket)}}=\underbrace{\overbrace{\sum_{s_z1,\ldots.s_zN}}^{\text{discrete lables}}\overbrace{\int_{R_N}d^3rN\ldots\int_{R_1}d^3r_1}^{\text{continuous lables}}}_{\text{adding up}}\underbrace{\Psi\left(r_1,\ldots\ldots rN,s_{z1},\ldots.s_zN\right)}_{\text{wave function (component of state vector along basis state)}}\underbrace{\left|r_1,\ldots.r_N,s_z1,\ldots.s_z\ N\right\rangle}_{\text{basis state (basis ket)}}.$$

For identical particles, symmetry requirements apply to both position and spin arguments of the wave function so it has the overall correct symmetry.

The formulae for the inner products are integrals over all coordinates or momenta and sums over all spin quantum numbers. For the general case of N particles with spin in 3d,

$$(\Psi_1,\Psi_2)=\sum_{s_{zN}}\cdots\sum_{s_{z2}}\sum_{s_{z1}}\int_{\text{all space}}d^3\mathbf{r}_1\int_{\text{all space}}d^3\mathbf{r}_2\cdots\int_{\text{all space}}d^3\mathbf{r}_N\Psi_1^*\left(\mathbf{r}_1\cdots\mathbf{r}_N,s_{z1}\cdots s_{zN},t\right)\Psi_2\left(\mathbf{r}_1\cdots\mathbf{r}_N,s_{z1}\cdots s_{zN},t\right)$$

this is altogether N three-dimensional volume integrals and N sums over the spins. The differential volume elements $d^3\mathbf{r}_i$ are also written "dV_i" or "$dx_i\,dy_i\,dz_i$".

The multidimensional Fourier transforms of the position or position–spin space wave functions yields momentum or momentum–spin space wave functions.

Probability Interpretation

For the general case of N particles with spin in 3d, if Ψ is interpreted as a probability amplitude, the probability density is,

$$\rho\left(\mathbf{r}_1\cdots\mathbf{r}_N,s_{z1}\cdots s_{zN},t\right)=\left|\Psi\left(\mathbf{r}_1\cdots\mathbf{r}_N,s_{z1}\cdots s_{zN},t\right)\right|^2$$

and the probability that particle 1 is in region R_1 with spin $s_{z1}=m_1$ *and* particle 2 is in region R_2 with spin $s_{z2}=m_2$ etc. at time t is the integral of the probability density over these regions and evaluated at these spin numbers:

$$P_{\mathbf{r}_1\in R_1,s_{z1}=m_1,\ldots,\mathbf{r}_N\in R_N,s_{zN}=m_N}(t)=\int_{R_1}d^3\mathbf{r}_1\int_{R_2}d^3\mathbf{r}_2\cdots\int_{R_N}d^3\mathbf{r}_N\left|\Psi\left(\mathbf{r}_1\cdots\mathbf{r}_N,m_1\cdots m_N,t\right)\right|^2.$$

Time Dependence

For systems in time-independent potentials, the wave function can always be written as a function of the degrees of freedom multiplied by a time-dependent phase factor, the form of which is given by the Schrödinger equation. For N particles, considering their positions only and suppressing other degrees of freedom,

$$\Psi(\mathbf{r}_1, \mathbf{r}_2, \ldots, \mathbf{r}_N, t) = e^{-iEt/\hbar}\, \psi(\mathbf{r}_1, \mathbf{r}_2, \ldots, \mathbf{r}_N),$$

where E is the energy eigenvalue of the system corresponding to the eigenstate Ψ. Wave functions of this form are called stationary states.

The time dependence of the quantum state and the operators can be placed according to unitary transformations on the operators and states. For any quantum state $|\Psi\rangle$ and operator O, in the Schrödinger picture $|\Psi(t)\rangle$ changes with time according to the Schrödinger equation while O is constant. In the Heisenberg picture it is the other way round, $|\Psi\rangle$ is constant while $O(t)$ evolves with time according to the Heisenberg equation of motion. The Dirac (or interaction) picture is intermediate, time dependence is places in both operators and states which evolve according to equations of motion. It is useful primarily in computing S-matrix elements.

Non-relativistic Examples

The following are solutions to the Schrödinger equation for one nonrelativistic spinless particle.

Finite Potential Barrier

One of most prominent features of the wave mechanics is a possibility for a particle to reach a location with a prohibitive (in classical mechanics) force potential. A common model is the "potential barrier", the one-dimensional case has the potential:

$$V(x) = \begin{cases} V_0 & |x| < a \\ 0 & |x| \geq L \end{cases}$$

and the steady-state solutions to the wave equation have the form (for some constants k, κ),

$$\Psi(x) = \begin{cases} A_r e^{ikx} + A_l e^{-ikx} & x < -a, \\ B_r e^{\kappa x} + B_l e^{-\kappa x} & |x| \leq a, \\ C_r e^{ikx} + C_l e^{-ikx} & x > a. \end{cases}$$

Note that these wave functions are not normalized.

The standard interpretation of this is as a stream of particles being fired at the step from the left (the direction of negative x): setting $A_r = 1$ corresponds to firing particles singly; the terms containing A_r and C_r signify motion to the right, while A_l and C_l – to the left. Under this beam interpretation, put $C_l = 0$ since no particles are coming from the right. By applying the continuity of wave functions and their derivatives at the boundaries, it is hence possible to determine the constants above.

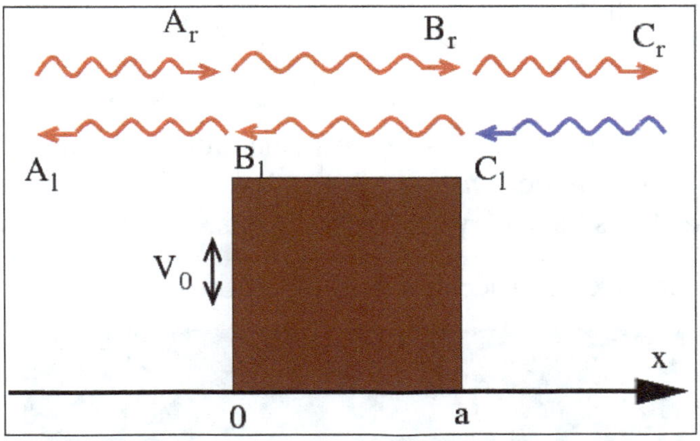

Scattering at a finite potential barrier of height V_0. The amplitudes and direction of left and right moving waves are indicated. In red, those waves used for the derivation of the reflection and transmission amplitude. $E > V_0$ for this illustration.

In a semiconductor crystallite whose radius is smaller than the size of its exciton Bohr radius, the excitons are squeezed, leading to quantum confinement. The energy levels can then be modeled using the particle in a box model in which the energy of different states is dependent on the length of the box.

Quantum Harmonic Oscillator

The wave functions for the quantum harmonic oscillator can be expressed in terms of Hermite polynomials H_n, they are:

$$\Psi_n(x) = \sqrt{\frac{1}{2^n\, n!}} \cdot \left(\frac{m\omega}{\pi\hbar}\right)^{1/4} \cdot e^{-\frac{m\omega x^2}{2\hbar}} \cdot H_n\left(\sqrt{\frac{m\omega}{\hbar}}x\right)$$

where n = 0,1,2,....

Hydrogen Atom

The wave functions of an electron in a Hydrogen atom are expressed in terms of spherical harmonics and generalized Laguerre polynomials.

It is convenient to use spherical coordinates, and the wave function can be separated into functions of each coordinate,

$$\Psi_{n\ell m}(r,\theta,\phi) = R(r)\; Y_\ell^m(\theta,\phi)$$

where R are radial functions and $Y^m{}_\ell(\theta,\varphi)$ are spherical harmonics of degree ℓ and order m. This is the only atom for which the Schrödinger equation has been solved exactly. Multi-electron atoms require approximative methods. The family of solutions is:

$$\Psi_{n\ell m}(r,\theta,\phi)=\sqrt{\left(\frac{2}{na_0}\right)^3\frac{(n-\ell-1)!}{2n[(n+\ell)!]}}e^{-r/na_0}\left(\frac{2r}{na_0}\right)^\ell L_{n-\ell-1}^{2\ell+1}\left(\frac{2r}{na_0}\right)\cdot Y_\ell^m(\theta,\phi)$$

where $a_0 = 4\pi\varepsilon_0\hbar^2/m_e e^2$ is the Bohr radius, $L^{2\ell+1}_{n-\ell-1}$ are the generalized Laguerre polynomials of degree $n - \ell - 1$, $n = 1, 2, \ldots$ is the principal quantum number, $\ell = 0, 1, \ldots n - 1$ the azimuthal quantum number, $m = -\ell, -\ell + 1, \ldots, \ell - 1, \ell$ the magnetic quantum number. Hydrogen-like atoms have very similar solutions.

This solution does not take into account the spin of the electron.

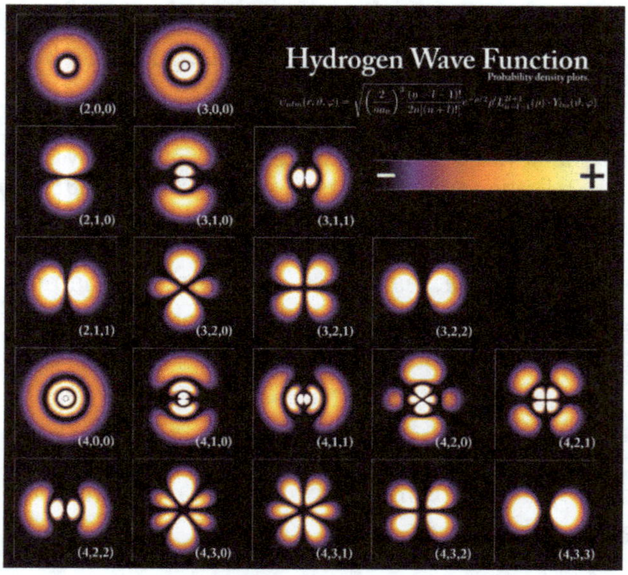

The electron probability density for the first few hydrogen atom electron orbitals shown as cross-sections. These orbitals form an orthonormal basis for the wave function of the electron. Different orbitals are depicted with different scale.

In the figure of the hydrogen orbitals, the 19 sub-images are images of wave functions in position space (their norm squared). The wave functions represent the abstract state characterized by the triple of quantum numbers (n, l, m), in the lower right of each image. These are the principal quantum number, the orbital angular momentum quantum number, and the magnetic quantum number. Together with one spin-projection quantum number of the electron, this is a complete set of observables.

The figure can serve to illustrate some further properties of the function spaces of wave functions:

- In this case, the wave functions are square integrable. One can initially take the function space as the space of square integrable functions, usually denoted L^2.

- The displayed functions are solutions to the Schrödinger equation. Obviously, not every function in L^2 satisfies the Schrödinger equation for the hydrogen atom. The function space is thus a subspace of L^2.

- The displayed functions form part of a basis for the function space. To each triple (n, l, m), there corresponds a basis wave function. If spin is taken into account, there are two basis functions for each triple. The function space thus has a countable basis.

- The basis functions are mutually orthonormal.

Wave Functions and Function Spaces

The concept of function spaces enters naturally in the discussion about wave functions. A function space is a set of functions, usually with some defining requirements on the functions (in the present case that they are square integrable), sometimes with an algebraic structure on the set (in the present case a vector space structure with an inner product), together with a topology on the set. The latter will sparsely be used here, it is only needed to obtain a precise definition of what it means for a subset of a function space to be closed. It will be concluded below that the function space of wave functions is a Hilbert space. This observation is the foundation of the predominant mathematical formulation of quantum mechanics.

Vector Space Structure

A wave function is an element of a function space partly characterized by the following concrete and abstract descriptions:

- The Schrödinger equation is linear. This means that the solutions to it, wave functions, can be added and multiplied by scalars to form a new solution. The set of solutions to the Schrödinger equation is a vector space.

- The superposition principle of quantum mechanics. If Ψ and Φ are two states in the abstract space of states of a quantum mechanical system, and a and b are any two complex numbers, then $a\Psi + b\Phi$ is a valid state as well. (Whether the null vector counts as a valid state ("no system present") is a matter of definition. The null vector does *not* at any rate describe the vacuum state in quantum field theory.) The set of allowable states is a vector space.

This similarity is of course not accidental. There are also a distinctions between the spaces.

Representations

Basic states are characterized by a set of quantum numbers. This is a set of eigenvalues of a maximal set of commuting observables. Physical observables are represented by linear operators, also called observables, on the vectors space. Maximality means that there can be added to the set no further algebraically independent observables that commute with the ones already present. A choice of such a set may be called a choice of representation.

- It is a postulate of quantum mechanics that a physically observable quantity of a system, such as position, momentum, or spin, is represented by a linear Hermitian operator on the state space. The possible outcomes of measurement of the quantity are the eigenvalues of the operator. At a deeper level, most observables, perhaps all, arise as generators of symmetries.
- The physical interpretation is that such a set represents what can – in theory – be simultaneously be measured with arbitrary precision. The Heisenberg uncertainty relation prohibits simultaneous exact measurements of two non-commuting observables.
- The set is non-unique. It may for a one-particle system, for example, be position and spin z-projection, (x, S_z), or it may be momentum and spin y-projection, (p, S_y). In this case, the operator corresponding to position (a multiplication operator in the position representation) and the operator corresponding to momentum (a differential operator in the position representation) do not commute.
- Once a representation is chosen, there is still arbitrariness. It remains to choose a coordinate system. This may, for example, correspond to a choice of x, y- and z-axis, or a choice of curvilinear coordinates as exemplified by the spherical coordinates used for the Hydrogen atomic wave functions. This final choice also fixes a basis in abstract Hilbert space. The basic states are labeled by the quantum numbers corresponding to the maximal set of commuting observables and an appropriate coordinate system.

The abstract states are "abstract" only in that an arbitrary choice necessary for a particular *explicit* description of it is not given. This is the same as saying that no choice of maximal set of commuting observables has been given. This is analogous to a vector space without a specified basis. Wave functions corresponding to a state are accordingly not unique. This non-uniqueness reflects the non-uniqueness in the choice of a maximal set of commuting observables. For one spin particle in one dimension, to a particular state there corresponds two wave functions, $\Psi(x, S_z)$ and $\Psi(p, S_y)$, both describing the *same* state:

- For each choice of maximal commuting sets of observables for the abstract state space, there is a corresponding representation that is associated to a function space of wave functions.
- Between all these different function spaces and the abstract state space, there are one-to-one correspondences (here disregarding normalization and unobservable phase factors), the common denominator here being a particular

abstract state. The relationship between the momentum and position space wave functions, for instance, describing the same state is the Fourier transform.

Each choice of representation should be thought of as specifying a unique function space in which wave functions corresponding to that choice of representation lives. This distinction is best kept, even if one could argue that two such function spaces are mathematically equal, e.g. being the set of square integrable functions. One can then think of the function spaces as two distinct copies of that set.

Inner Product

There is an additional algebraic structure on the vector spaces of wave functions and the abstract state space:

- Physically, different wave functions are interpreted to overlap to some degree. A system in a state Ψ that does *not* overlap with a state Φ cannot be found to be in the state Φ upon measurement. But if Φ_1, Φ_2, ... overlap Ψ to *some* degree, there is a chance that measurement of a system described by Ψ will be found in states Φ_1, Φ_2, ... Also selection rules are observed apply. These are usually formulated in the preservation of some quantum numbers. This means that certain processes allowable from some perspectives (e.g. energy and momentum conservation) do not occur because the initial and final *total* wave functions don't overlap.

- Mathematically, it turns out that solutions to the Schrödinger equation for particular potentials are orthogonal in some manner, this is usually described by an integral.

  $$\int \Psi_m^* \Psi_n w dV = \delta_{nm},$$

  where *m*, *n* are (sets of) indices (quantum numbers) labeling different solutions, the strictly positive function *w* is called a weight function, and δ_{mn} is the Kronecker delta. The integration is taken over all of the relevant space.

This motivates the introduction of an inner product on the vector space of abstract quantum states, compatible with the mathematical observations above when passing to a representation. It is denoted (Ψ, Φ), or in the Bra–ket notation $\langle\Psi|\Phi\rangle$. It yields a complex number. With the inner product, the function space is an inner product space. The explicit appearance of the inner product (usually an integral or a sum of integrals) depends on the choice of representation, but the complex number (Ψ, Φ) does not. Much of the physical interpretation of quantum mechanics stems from the Born rule. It states that the probability *p* of finding upon measurement the state Φ given the system is in the state Ψ is:

$$p = |(\Phi, \Psi)|^2,$$

where Φ and Ψ are assumed normalized. Consider a scattering experiment. In quantum field theory, if Φ_{out} describes a state in the "distant future" (an "out state") after interactions between scattering particles have ceased, and Ψ_{in} an "in state" in the "distant past", then the quantities (Φ_{out}, Ψ_{in}), with Φ_{out} and Ψ_{in} varying over a complete set of in states and out states respectively, is called the S-matrix or scattering matrix. Knowledge of it is, effectively, having *solved* the theory at hand, at least as far as predictions go. Measurable quantities such as decay rates and scattering cross sections are calculable from the S-matrix.

Hilbert Space

The above observations encapsulate the essence of the function spaces of which wave functions are elements. However, the description is not yet complete. There is a further technical requirement on the function space, that of completeness, that allows one to take limits of sequences in the function space, and be ensured that, if the limit exists, it is an element of the function space. A complete inner product space is called a Hilbert space. The property of completeness is crucial in advanced treatments and applications of quantum mechanics. For instance, the existence of projection operators or orthogonal projections relies on the completeness of the space. These projection operators, in turn, are essential for the statement and proof of many useful theorems, e.g. the spectral theorem. It is not very important in introductory quantum mechanics, and technical details and links may be found in footnotes like the one that follows. The space L^2 is a Hilbert space, with inner product presented later. The function space of the example of the figure is a subspace of L^2. A subspace of a Hilbert space is a Hilbert space if it is closed.

The set of all possible normalizable wave functions for a system with a particular choice of basis, together with the null vector, constitute a Hilbert space.

Not all functions of interest are elements of some Hilbert space, say L^2. The most glaring example is the set of functions $e^{2\pi i p \cdot x/h}$. These are plane wave solutions of the Schrödinger equation for a free particle, but are not normalizable, hence not in L^2. But they are nonetheless fundamental for the description. One can, using them, express functions that *are* normalizable using wave packets. They are, in a sense, a basis (but not a Hilbert space basis, nor a Hamel basis) in which wave functions of interest can be expressed. There is also the artifact "normalization to a delta function" that is frequently employed for notational convenience. The delta functions themselves aren't square integrable either.

The description of the function space containing the wave functions is mostly mathematically motivated. The function spaces are, due to completeness, very *large* in a certain sense. Not all functions are realistic descriptions of any physical system. For instance, in the function space L^2 one can find the function that takes on the value 0 for all rational numbers and -*i* for the irrationals in the interval [0, 1]. This *is* square integrable, but can hardly represent a physical state.

Common Hilbert Spaces

While the space of solutions as a whole is a Hilbert space there are many other Hilbert spaces that commonly occur as ingredients:

- Square integrable complex valued functions on the interval $[0, 2\pi]$. The set $\{e^{int}/2\pi, n \in \mathbb{Z}\}$ is a Hilbert space basis, i.e. a maximal orthonormal set.
- The Fourier transform takes functions in the above space to elements of $l^2(\mathbb{Z})$, the space of *square summable* functions $\mathbb{Z} \to \mathbb{C}$. The latter space is a Hilbert space and the Fourier transform is an isomorphism of Hilbert spaces. Its basis is $\{e_i, i \in \mathbb{Z}\}$ with $e_i(j) = \delta_{ij}, i, j \in \mathbb{Z}$.
- The most basic example of spanning polynomials is in the space of square integrable functions on the interval $[-1, 1]$ for which the Legendre polynomials is a Hilbert space basis (complete orthonormal set).
- The square integrable functions on the unit sphere S^2 is a Hilbert space. The basis functions in this case are the spherical harmonics. The Legendre polynomials are ingredients in the spherical harmonics. Most problems with rotational symmetry will have "the same" (known) solution with respect to that symmetry, so the original problem is reduced to a problem of lower dimensionality.
- The associated Laguerre polynomials appear in the hydrogenic wave function problem after factoring out the spherical harmonics. These span the Hilbert space of square integrable functions on the semi-infinite interval $[0, \infty)$.

More generally, one may consider a unified treatment of all second order polynomial solutions to the Sturm–Liouville equations in the setting of Hilbert space. These include the Legendre and Laguerre polynomials as well as Chebyshev polynomials, Jacobi polynomials and Hermite polynomials. All of these actually appear in physical problems, the latter ones in the harmonic oscillator, and what is otherwise a bewildering maze of properties of special functions becomes an organized body of facts.

There occurs also finite-dimensional Hilbert spaces. The space $\mathbb{C}^n$ is a Hilbert space of dimension n. The inner product is the standard inner product on these spaces. In it, the "spin part" of a single particle wave function resides:

- In the non-relativistic description of an electron one has $n = 2$ and the total wave function is a solution of the Pauli equation.
- In the corresponding relativistic treatment, $n = 4$ and the wave function solves the Dirac equation.

With more particles, the situations is more complicated. One has to employ tensor products and use representation theory of the symmetry groups involved (the rotation group and the Lorentz group respectively) to extract from the tensor product the spaces in which the (total) spin wave functions reside. (Further problems arise in the relativistic case unless the particles are free.) Corresponding remarks apply to the concept of isospin, for which the symmetry group is SU(2). The models of the nuclear forces of the sixties used the symmetry group SU(3). In this case, as well, the part of the wave functions corresponding to the inner symmetries reside in some $\mathbb{C}^n$ or subspaces of tensor products of such spaces:

- In quantum field theory the underlying Hilbert space is Fock space. It is built from free single-particle states, i.e. wave functions when a representation is chosen, and can accommodate any finite, not necessarily constant in time, number of particles. The interesting (or rather the *tractable*) dynamics lies not in the wave functions but in the field operators that are operators acting on Fock space. Thus the Heisenberg picture is the most common choice (constant states, time varying operators).

Due to the infinite-dimensional nature of the system, the appropriate mathematical tools are objects of study in functional analysis.

Simplified Description

Not all introductory textbooks take the long route and introduce the full Hilbert space machinery, but the focus is on the non-relativistic Schrödinger equation in position representation for certain standard potentials. The following constraints on the wave function are sometimes explicitly formulated for the calculations and physical interpretation to make sense:

- The wave function must be square integrable. This is motivated by the Copenhagen interpretation of the wave function as a probability amplitude.
- It must be everywhere continuous and everywhere continuously differentiable. This is motivated by the appearance of the Schrödinger equation for most physically reasonable potentials.

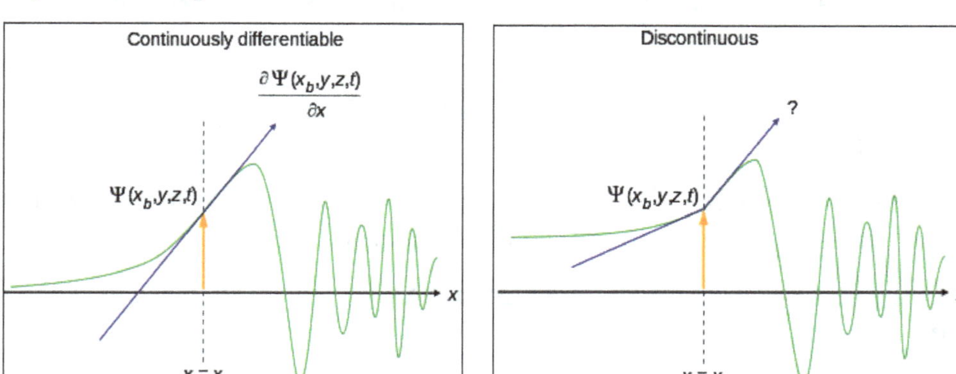

Continuity of the wave function and its first spatial derivative
(in the x direction, y and z coordinates not shown), at some time t.

It is possible to relax these conditions somewhat for special purposes. If these requirements are not met, it is not possible to interpret the wave function as a probability amplitude.

This does not alter the structure of the Hilbert space that these particular wave functions inhabit, but it should be pointed out that the subspace of the square-integrable functions L^2, which is a Hilbert space, satisfying the second requirement *is not closed* in L^2, hence not a Hilbert space in itself. The functions that does not meet the requirements are still needed for both technical and practical reasons.

Wave Functions and Abstract State Space

As has been demonstrated, the set of all possible wave functions in some representation for a system constitute an in general infinite-dimensional Hilbert space. Due to the multiple possible choices of representation basis, these Hilbert spaces are not unique. One therefore talks about an abstract Hilbert space, state space, where the choice of representation and basis is left undetermined. Specifically, each state is represented as an abstract vector in state space. A quantum state $|\Psi\rangle$ in any representation is generally expressed as a vector,

$$|\Psi\rangle = \sum_{\alpha} \int d^m \omega \, \Psi(\alpha,\omega,t) | \alpha,\omega\rangle$$

where,

- $|\alpha, \omega\rangle$ the basis vectors of the chosen representation.
- $d^m\omega = d\omega_1 d\omega_2 ... d\omega_m$ a "differential volume element" in the continuous degrees of freedom.
- $\Psi(\alpha, \omega, t)$ a component of the vector $|\Psi\rangle$, called the wave function of the system.
- $\alpha = (\alpha_1, \alpha_2, ..., \alpha_n)$ dimensionless discrete quantum numbers.
- $\omega = (\omega_1, \omega_2, ..., \omega_m)$ continuous variables (not necessarily dimensionless).

These quantum numbers index the components of the state vector. More, all α are in an n-dimensional set $A = A_1 \times A_2 \times ... A_n$ where each A_i is the set of allowed values for α_i; all ω are in an m-dimensional "volume" $\Omega \subseteq \mathbb{R}^m$ where $\Omega = \Omega_1 \times \Omega_2 \times ... \Omega_m$ and each $\Omega_i \subseteq \mathbb{R}$ is the set of allowed values for ω_i, a subset of the real numbers $\mathbb{R}$. For generality n and m are not necessarily equal.

Example:

1. For a single particle in 3d with spin s, neglecting other degrees of freedom, using Cartesian coordinates, we could take $\alpha = (s_z)$ for the spin quantum number of the particle along the z direction, and $\omega = (x, y, z)$ for the particle's position coordinates. Here

$A = \{-s, -s+1, ..., s-1, s\}$ is the set of allowed spin quantum numbers and $\Omega = \mathbb{R}^3$ is the set of all possible particle positions throughout 3d position space.

2. An alternative choice is $\alpha = (s_y)$ for the spin quantum number along the y direction and $\omega = (p_x, p_y, p_z)$ for the particle's momentum components. In this case A and Ω are the same as before.

The probability density of finding the system at time t at state $|\alpha, \omega\rangle$ is:

$$\rho_{\alpha,\omega}(t) = |\Psi(\alpha,\omega,t)|^2$$

The probability of finding system with α in some or all possible discrete-variable configurations, $D \subseteq A$, and ω in some or all possible continuous-variable configurations, $C \subseteq \Omega$, is the sum and integral over the density:

$$P(t) = \sum_{\alpha \in D} \int_C \rho_{\alpha,\omega}(t)\ d^m\omega$$

Since the sum of all probabilities must be 1, the normalization condition:

$$1 = \sum_{\alpha \in A} \int_\Omega \rho_{\alpha,\omega}(t)\ d^m\omega$$

must hold at all times during the evolution of the system.

The normalization condition requires $\rho\ d^m\omega$ to be dimensionless, by dimensional analysis Ψ must have the same units as $(\omega_1\omega_2...\omega_m)^{-1/2}$.

OPERATOR

In physics, an operator is a function over a space of physical states to another space of physical states. The simplest example of the utility of operators is the study of symmetry (which makes the concept of a group useful in this context). Because of this, they are very useful tools in classical mechanics. Operators are even more important in quantum mechanics, where they form an intrinsic part of the formulation of the theory.

Operators in Quantum Mechanics

The mathematical formulation of quantum mechanics (QM) is built upon the concept of an operator.

The wavefunction represents the probability amplitude of finding the system in that state. The terms "wavefunction" and "state" in QM context are usually used interchangeably.

Physical pure states in quantum mechanics are represented as unit-norm vectors (probabilities are normalized to one) in a special complex Hilbert space. Time evolution in this vector space is given by the application of the evolution operator.

Any observable, i.e., any quantity which can be measured in a physical experiment, should be associated with a self-adjoint linear operator. The operators must yield real eigenvalues, since they are values which may come up as the result of the experiment. Although traditionally physicists associated real eigenvalues with Hermiticity, in 1998 physicists realized that there also exist non-Hermitian operators with all-real spectra; namely, parity-time (PT) symmetric operators. For Hermitian operators, the probability of each eigenvalue is related to the projection of the physical state on the subspace related to that eigenvalue.

In the wave mechanics formulation of QM, the wavefunction varies with space and time, or equivalently momentum and time, so observables are differential operators.

In the matrix mechanics formulation, the norm of the physical state should stay fixed, so the evolution operator should be unitary, and the operators can be represented as matrices. Any other symmetry, mapping a physical state into another, should keep this restriction.

Wavefunction

The wavefunction must be square-integrable, meaning:

$$\iiint_{\mathbb{R}^3} |\psi(\mathbf{r})|^2 \, \mathrm{d}^3\mathbf{r} = \iiint_{\mathbb{R}^3} \psi(\mathbf{r})^* \psi(\mathbf{r}) \mathrm{d}^3\mathbf{r} < \infty$$

and normalizable, so that:

$$\iiint_{\mathbb{R}^3} |\psi(\mathbf{r})|^2 \, \mathrm{d}^3\mathbf{r} = 1$$

Two cases of eigenstates (and eigenvalues) are:

- For discrete eigenstates $\psi_i\rangle$ forming a discrete basis, so the state is a sum:

  $$|\psi\rangle = \sum_i c_i \, |\phi_i\rangle$$

  where c_i are complex numbers such that $|c_i|^2 = c_i^* c_i$ = probability of measuring the state $|\phi_i\rangle$, and has the corresponding set of eigenvalues a_i is also discrete - either finite or countably infinite.

- For a continuum of eigenstates $\psi_i\rangle$ angle forming a continuous basis, so the state is an integral:

  $$|\psi\rangle = \int c(\phi) \mathrm{d}\phi \, |\phi\rangle$$

where $c(\varphi)$ is a complex function such that $|c(\varphi)|^2 = c(\varphi)^* c(\varphi)$ = probability of measuring the state $|\phi\rangle$, there is an uncountably infinite set of eigenvalues a.

Linear Operators in Wave Mechanics

Let ψ be the wavefunction for a quantum system, and $\hat{A}$ be any linear operator for some observable A (such as position, momentum, energy, angular momentum etc.), then,

$$\hat{A}\psi = a\psi,$$

where:

- a is the eigenvalue of the operator, corresponding to the measured value of the observable, i.e. observable A has a measured value a.
- ψ is the eigenfunction of $\hat{A}$ if this relation holds.

If ψ is an eigenfunction of a given operator A, then a definite quantity (the eigenvalue a) will be observed if a measurement of the observable A is made on the state ψ. Conversely, if ψ is not an eigenfunction of A, then it has no eigenvalue for A, and the observable does not have a single definite value in that case. Instead, measurements of the observable A will yield each eigenvalue with a certain probability (related to the decomposition of ψ relative to the orthonormal eigenbasis of A).

In bra–ket notation the above can be written;

$$\hat{A}\psi = \hat{A}\psi(\mathbf{r}) = \hat{A}\langle \mathbf{r} | \psi \rangle = \langle \mathbf{r} | \hat{A} | \psi \rangle$$
$$a\psi = a\psi(\mathbf{r}) = a\langle \mathbf{r} | \psi \rangle = \langle \mathbf{r} | a | \psi \rangle$$

in which case $|\psi\rangle$ is an eigenvector, or eigenket.

Due to linearity, vectors can be defined in any number of dimensions, as each component of the vector acts on the function separately. One mathematical example is the del operator, which is itself a vector.

An operator in n-dimensional space can be written:

$$\hat{\mathbf{A}} = \sum_{j=1}^{n} \mathbf{e}_j \hat{A}_j$$

where $\mathbf{e}_j$ are basis vectors corresponding to each component operator A_j. Each component will yield a corresponding eigenvalue. Acting this on the wave function ψ:

$$\hat{\mathbf{A}}\psi = \left(\sum_{j=1}^{n} \mathbf{e}_j \hat{A}_j \right) \psi = \sum_{j=1}^{n} \left(\mathbf{e}_j \hat{A}_j \psi \right) = \sum_{j=1}^{n} \left(\mathbf{e}_j a_j \psi \right)$$

in which,

$$\hat{A}_j \psi = a_j \psi.$$

In bra–ket notation:

$$\hat{\mathbf{A}}\psi = \hat{\mathbf{A}}\psi(\mathbf{r}) = \hat{\mathbf{A}}\langle \mathbf{r} | \psi \rangle = \langle \mathbf{r} | \hat{\mathbf{A}} | \psi \rangle$$

$$\left(\sum_{j=1}^{n} \mathbf{e}_j \hat{A}_j\right)\psi = \left(\sum_{j=1}^{n} \mathbf{e}_j \hat{A}_j\right)\psi(\mathbf{r}) = \left(\sum_{j=1}^{n} \mathbf{e}_j \hat{A}_j\right)\langle \mathbf{r} | \psi \rangle = \left\langle \mathbf{r} \left| \sum_{j=1}^{n} \mathbf{e}_j \hat{A}_j \right| \psi \right\rangle$$

Commutation of Operators on Ψ

If two observables *A* and *B* have linear operators $\hat{\mathrm{A}}$ and $\hat{\mathrm{B}}$, the commutator is defined by,

$$\left[\hat{A}, \hat{B}\right] = \hat{A}\hat{B} - \hat{B}\hat{A}$$

The commutator is itself a (composite) operator. Acting the commutator on ψ gives:

$$\left[\hat{A}, \hat{B}\right]\psi = \hat{A}\hat{B}\psi - \hat{B}\hat{A}\psi.$$

If ψ is an eigenfunction with eigenvalues a and b for observables A and B respectively, and if the operators commute:

$$\left[\hat{A}, \hat{B}\right]\psi = 0,$$

then the observables *A* and *B* can be measured simultaneously with infinite precision i.e. uncertainties $\Delta A = 0$, $\Delta B = 0$ simultaneously. ψ is then said to be the simultaneous eigenfunction of A and B. To illustrate this:

$$\begin{aligned}\left[\hat{A}, \hat{B}\right]\psi &= \hat{A}\hat{B}\psi - \hat{B}\hat{A}\psi \\ &= a(b\psi) - b(a\psi) \\ &= 0.\end{aligned}$$

It shows that measurement of A and B does not cause any shift of state i.e. initial and final states are same (no disturbance due to measurement). Suppose we measure A to get value a. We then measure B to get the value b. We measure A again. We still get the same value a. Clearly the state (ψ) of the system is not destroyed and so we are able to measure A and B simultaneously with infinite precision.

If the operators do not commute:

$$\left[\hat{A}, \hat{B}\right]\psi \neq 0,$$

they can't be prepared simultaneously to arbitrary precision, and there is an uncertainty relation between the observables,

$$\Delta A \Delta B \geq \left|\frac{\langle [A, B] \rangle}{2}\right|$$

even if ψ is an eigenfunction the above relation holds.. Notable pairs are position and momentum, and energy and time - uncertainty relations, and the angular momenta (spin, orbital and total) about any two orthogonal axes (such as L_x and L_y, or s_y and s_z etc.).

Expectation Values of Operators on Ψ

The expectation value (equivalently the average or mean value) is the average measurement of an observable, for particle in region R. The expectation value $\langle \hat{A} \rangle$ of the operator $\hat{A}$ is calculated from:

$$\langle \hat{A} \rangle = \int_R \psi^*(\mathbf{r}) \hat{A} \psi(\mathbf{r}) \mathrm{d}^3\mathbf{r} = \langle \psi | \hat{A} | \psi \rangle$$

This can be generalized to any function F of an operator:

$$\langle F(\hat{A}) \rangle = \int_R \psi(\mathbf{r})^* \left[F(\hat{A}) \psi(\mathbf{r}) \right] \mathrm{d}^3\mathbf{r} = \langle \psi | F(\hat{A}) | \psi \rangle,$$

An example of F is the 2-fold action of A on ψ, i.e. squaring an operator or doing it twice:

$$F(\hat{A}) = \hat{A}^2$$

$$\Rightarrow \langle \hat{A}^2 \rangle = \int_R \psi^*(\mathbf{r}) \hat{A}^2 \psi(\mathbf{r}) \mathrm{d}^3\mathbf{r} = \langle \psi | \hat{A}^2 | \psi \rangle$$

Hermitian Operators

The definition of a Hermitian operator is:

$$\hat{A} = \hat{A}^\dagger$$

Following from this, in bra–ket notation:

$$\langle \phi_i | \hat{A} | \phi_j \rangle = \langle \phi_j | \hat{A} | \phi_i \rangle^*.$$

Important properties of Hermitian operators include:

- Real eigenvalues.
- Eigenvectors with different eigenvalues are orthogonal.
- Eigenvectors can be chosen to be a complete orthonormal basis.

Operators in Matrix Mechanics

An operator can be written in matrix form to map one basis vector to another. Since the operators are linear, the matrix is a linear transformation (aka transition matrix) between bases. Each basis element ϕ_j can be connected to another, by the expression:

$$A_{ij} = \langle \phi_i | \hat{A} | \phi_j \rangle,$$

which is a matrix element:

$$\hat{A} = \begin{pmatrix} A_{11} & A_{12} & \cdots & A_{1n} \\ A_{21} & A_{22} & \cdots & A_{2n} \\ \vdots & \vdots & \ddots & \vdots \\ A_{n1} & A_{n2} & \cdots & A_{nn} \end{pmatrix}$$

A further property of a Hermitian operator is that eigenfunctions corresponding to different eigenvalues are orthogonal. In matrix form, operators allow real eigenvalues to be found, corresponding to measurements. Orthogonality allows a suitable basis set of vectors to represent the state of the quantum system. The eigenvalues of the operator are also evaluated in the same way as for the square matrix, by solving the characteristic polynomial:

$$\det\left(\hat{A} - a\hat{I}\right) = 0,$$

where I is the n × n identity matrix, as an operator it corresponds to the identity operator. For a discrete basis:

$$\hat{I} = \sum_i |\phi_i\rangle\langle\phi_i|$$

while for a continuous basis:

$$\hat{I} = \int |\phi\rangle\langle\phi|\, d\phi$$

Inverse of an Operator

A non-singular operator $\hat{A}$ has an inverse $\hat{A}^{-1}$ defined by:

$$\hat{A}\hat{A}^{-1} = \hat{A}^{-1}\hat{A} = \hat{I}$$

If an operator has no inverse, it is a singular operator. In a finite-dimensional space, an operator is non-singular if and only if its determinant is nonzero:

$$\det(\hat{A}) \neq 0$$

and hence the determinant is zero for a singular operator.

Table of QM Operators

The operators used in quantum mechanics are collected in the table below. The boldface vectors with circumflexes are not unit vectors, they are 3-vector operators; all three spatial components taken together.

Operator (common name/s)	Cartesian component	General definition	SI unit	Dimension
Position	$\hat{x}=x$ $\hat{y}=y$ $\hat{z}=z$	$\hat{\mathbf{r}}=\mathbf{r}$	m	[L]
Momentum	General $\hat{p}_x=-i\hbar\frac{\partial}{\partial x}$ $\hat{p}_y=-i\hbar\frac{\partial}{\partial y}$ $\hat{p}_z=-i\hbar\frac{\partial}{\partial z}$	General $\hat{\mathbf{p}}=-i\hbar\nabla$	$\mathrm{J\,s\,m^{-1}}$ $=\mathrm{N\,s}$	$[\mathrm{M}]\,[\mathrm{L}]\,[\mathrm{T}]^{-1}$
	Electromagnetic field $\hat{p}_x=-i\hbar\frac{\partial}{\partial x}-qA_x$ $\hat{p}_y=-i\hbar\frac{\partial}{\partial y}-qA_y$ $\hat{p}_z=-i\hbar\frac{\partial}{\partial z}-qA_z$	Electromagnetic field (uses kinetic momentum, A = vector potential) $\hat{\mathbf{p}}=\hat{\mathbf{P}}-q\mathbf{A}$ $=-i\hbar\nabla-q\mathbf{A}$	$\mathrm{J\,s\,m^{-1}}$ $=\mathrm{N\,s}$	$[\mathrm{M}]\,[\mathrm{L}]\,[\mathrm{T}]^{-1}$
Kinetic energy	Translation $\hat{T}_x=-\frac{\hbar^2}{2m}\frac{\partial^2}{\partial x^2}$ $\hat{T}_y=-\frac{\hbar^2}{2m}\frac{\partial^2}{\partial y^2}$ $\hat{T}_z=-\frac{\hbar^2}{2m}\frac{\partial^2}{\partial z^2}$	$\hat{T}=\frac{\hat{\mathbf{p}}\cdot\hat{\mathbf{p}}}{2m}=\frac{(-i\hbar\nabla)\cdot(-i\hbar\nabla)}{2m}=\frac{-\hbar^2}{2m}\nabla^2$	J	$[\mathrm{M}]\,[\mathrm{L}]^2$ $[\mathrm{T}]^{-2}$
	Electromagnetic field $\hat{T}_x=\frac{1}{2m}\left(-i\hbar\frac{\partial}{\partial x}-qA_x\right)^2$ $\hat{T}_y=\frac{1}{2m}\left(-i\hbar\frac{\partial}{\partial y}-qA_y\right)^2$ $\hat{T}_z=\frac{1}{2m}\left(-i\hbar\frac{\partial}{\partial z}-qA_z\right)^2$	Electromagnetic field (A = vector potential) $\hat{T}=\frac{\hat{\mathbf{p}}\cdot\hat{\mathbf{p}}}{2m}$ $=\frac{1}{2m}(-i\hbar\nabla-q\mathbf{A})\cdot(-i\hbar\nabla-q\mathbf{A})$ $=\frac{1}{2m}(-i\hbar\nabla-q\mathbf{A})^2$	J	$[\mathrm{M}]\,[\mathrm{L}]^2$ $[\mathrm{T}]^{-2}$
	Rotation (I = moment of inertia) $\hat{T}_{xx}=\frac{\hat{J}_x^2}{2I_{xx}}$ $\hat{T}_{yy}=\frac{\hat{J}_y^2}{2I_{yy}}$ $\hat{T}_{zz}=\frac{\hat{J}_z^2}{2I_{zz}}$	Rotation $\hat{T}=\frac{\hat{\mathbf{J}}\cdot\hat{\mathbf{J}}}{2I}$	J	$[\mathrm{M}]\,[\mathrm{L}]^2$ $[\mathrm{T}]^{-2}$

Potential energy	N/A	$\hat{V} = V(\mathbf{r},t) = V$	J	[M] [L]^2 [T]$^{-2}$
Total energy	N/A	Time-dependent potential: $\hat{E} = i\hbar\frac{\partial}{\partial t}$ Time-independent: $\hat{E} = E$	J	[M] [L]^2 [T]$^{-2}$
Hamiltonian		$\hat{H} = \hat{T} + \hat{V}$ $= \frac{\hat{\mathbf{p}}\cdot\hat{\mathbf{p}}}{2m} + V$ $= \frac{\hat{p}^2}{2m} + V$	J	[M] [L]^2 [T]$^{-2}$
Angular momentum operator	$\hat{L}_x = -i\hbar\left(y\frac{\partial}{\partial z} - z\frac{\partial}{\partial y}\right)$ $\hat{L}_y = -i\hbar\left(z\frac{\partial}{\partial x} - x\frac{\partial}{\partial z}\right)$ $\hat{L}_z = -i\hbar\left(x\frac{\partial}{\partial y} - y\frac{\partial}{\partial x}\right)$	$\hat{\mathbf{L}} = \mathbf{r} \times -i\hbar\nabla$	J s = N s m	[M] [L]^2 [T]$^{-1}$
Spin angular momentum	$\hat{S}_x = \frac{\hbar}{2}\sigma_x$ $\hat{S}_y = \frac{\hbar}{2}\sigma_y$ $\hat{S}_z = \frac{\hbar}{2}\sigma_z$ where $\sigma_x = \begin{pmatrix} 0 & 1 \\ 1 & 0 \end{pmatrix}$ $\sigma_y = \begin{pmatrix} 0 & -i \\ i & 0 \end{pmatrix}$ $\sigma_z = \begin{pmatrix} 1 & 0 \\ 0 & -1 \end{pmatrix}$ are the pauli matrices for spin-½ particles.	$\hat{\mathbf{S}} = \frac{\hbar}{2}\sigma$ where σ is the vector whose components are the pauli matrices.	J s = N s m	[M] [L]^2 [T]$^{-1}$
Total angular momentum	$\hat{J}_x = \hat{L}_x + \hat{S}_x$ $\hat{J}_y = \hat{L}_y + \hat{S}_y$ $\hat{J}_z = \hat{L}_z + \hat{S}_z$	$\hat{\mathbf{J}} = \hat{\mathbf{L}} + \hat{\mathbf{S}}$ $= -i\hbar\mathbf{r} \times \nabla + \frac{\hbar}{2}\sigma$	J s = N s m	[M] [L]^2 [T]$^{-1}$

Transition dipole moment (electric)	$\hat{d}_x = q\hat{x}$ $\hat{d}_y = q\hat{y}$ $\hat{d}_z = q\hat{z}$	$\hat{\mathbf{d}} = q\hat{\mathbf{r}}$	C m	[I] [T] [L]

Examples of Applying Quantum Operators

The procedure for extracting information from a wave function is as follows. Consider the momentum *p* of a particle as an example. The momentum operator in position basis in one dimension is:

$$\hat{p} = -i\hbar \frac{\partial}{\partial x}$$

Letting this act on ψ we obtain:

$$\hat{p}\psi = -i\hbar \frac{\partial}{\partial x}\psi,$$

if ψ is an eigenfunction of $\hat{p}$, then the momentum eigenvalue p is the value of the particle's momentum, found by:

$$-i\hbar \frac{\partial}{\partial x}\psi = p\psi.$$

For three dimensions the momentum operator uses the nabla operator to become:

$$\hat{\mathbf{p}} = -i\hbar\nabla.$$

In Cartesian coordinates (using the standard Cartesian basis vectors $\mathbf{e}_x$, $\mathbf{e}_y$, $\mathbf{e}_z$) this can be written:

$$\mathbf{e}_x\hat{p}_x + \mathbf{e}_y\hat{p}_y + \mathbf{e}_z\hat{p}_z = -i\hbar\left(\mathbf{e}_x\frac{\partial}{\partial x} + \mathbf{e}_y\frac{\partial}{\partial y} + \mathbf{e}_z\frac{\partial}{\partial z}\right),$$

that is:

$$\hat{p}_x = -i\hbar\frac{\partial}{\partial x}, \quad \hat{p}_y = -i\hbar\frac{\partial}{\partial y}, \quad \hat{p}_z = -i\hbar\frac{\partial}{\partial z}$$

The process of finding eigenvalues is the same. Since this is a vector and operator equation, if ψ is an eigenfunction, then each component of the momentum operator will have an eigenvalue corresponding to that component of momentum. Acting $\hat{p}$ on ψ obtains:

$$\hat{p}_x\psi = -i\hbar\frac{\partial}{\partial x}\psi = p_x\psi$$

$$\hat{p}_y\psi = -i\hbar\frac{\partial}{\partial y}\psi = p_y\psi$$

$$\hat{p}_z\psi = -i\hbar\frac{\partial}{\partial z}\psi = p_z\psi$$

Hamiltonian

In quantum mechanics, a Hamiltonian is an operator corresponding to the sum of the kinetic energies plus the potential energies for all the particles in the system (this addition is the total energy of the system in most of the cases under analysis). It is usually denoted by H, but also $\check{H}$ or $\hat{H}$ to highlight its function as an operator. Its spectrum is the set of possible outcomes when one measures the total energy of a system. Because of its close relation to the time-evolution of a system, it is of fundamental importance in most formulations of quantum theory.

The Hamiltonian is named after William Rowan Hamilton, who created a revolutionary reformulation of Newtonian mechanics, now called Hamiltonian mechanics, which is also important in quantum physics.

The Hamiltonian is the sum of the kinetic energies of all the particles, plus the potential energy of the particles associated with the system. The expression of the Hamiltonian can take different forms and simplifications taken into account the concrete characteristics of the system under analysis: single or several particles in the system; interaction between particles; kind of potential energy; time varying potential or time independent one; etc.

Schrödinger Hamiltonian

One Particle

By analogy with classical mechanics, the Hamiltonian is commonly expressed as the sum of operators corresponding to the kinetic and potential energies of a system in the form:

$$\hat{H} = \hat{T} + \hat{V},$$

where,

$$\hat{V} = V = V(\mathbf{r}, t),$$

is the potential energy operator and,

$$\hat{T} = \frac{\hat{\mathbf{p}} \cdot \hat{\mathbf{p}}}{2m} = \frac{\hat{p}^2}{2m} = -\frac{\hbar^2}{2m}\nabla^2,$$

is the kinetic energy operator in which m is the mass of the particle, the dot denotes the dot product of vectors, and,

$$\hat{p} = -i\hbar\nabla,$$

is the momentum operator where an ∇ is the del operator. The dot product of ∇ with itself is the Laplacian ∇^2. In three dimensions using Cartesian coordinates the Laplace operator is:

$$\nabla^2 = \frac{\partial^2}{\partial x^2} + \frac{\partial^2}{\partial y^2} + \frac{\partial^2}{\partial z^2}$$

Although this is not the technical definition of the Hamiltonian in classical mechanics, it is the form it most commonly takes. Combining these together yields the familiar form used in the Schrödinger equation:

$$\begin{aligned} \hat{H} &= \hat{T} + \hat{V} \\ &= \frac{\hat{\mathbf{p}} \cdot \hat{\mathbf{p}}}{2m} + V(\mathbf{r}, t) \\ &= -\frac{\hbar^2}{2m}\nabla^2 + V(\mathbf{r}, t) \end{aligned}$$

which allows one to apply the Hamiltonian to systems described by a wave function $\Psi(\mathbf{r}, t)$. This is the approach commonly taken in introductory treatments of quantum mechanics, using the formalism of Schrödinger's wave mechanics.

One can also make substitutions to certain variables to fit specific cases, such as some involving electromagnetic fields.

Many Particles

The formalism can be extended to N particles:

$$\hat{H} = \sum_{n=1}^{N} \hat{T}_n + V$$

where,

$$V = V(\mathbf{r}_1, \mathbf{r}_2 \cdots \mathbf{r}_N, t),$$

is the potential energy function, now a function of the spatial configuration of the system and time (a particular set of spatial positions at some instant of time defines a configuration) and,

$$\hat{T}_n = \frac{\mathbf{p}_n \cdot \mathbf{p}_n}{2m_n},$$

is the kinetic energy operator of particle n, and Δ_n is the gradient for particle n, ∇_n^2 is the Laplacian for particle using the coordinates:

$$\nabla_n^2 = \frac{\partial^2}{\partial x_n^2} + \frac{\partial^2}{\partial y_n^2} + \frac{\partial^2}{\partial z_n^2},$$

Combining these yields the Schrödinger Hamiltonian for the N-particle case:

$$\hat{H} = \sum_{n=1}^{N} \hat{T}_n + V$$

$$= \sum_{n=1}^{N} \frac{\hat{\mathbf{p}}_n \cdot \hat{\mathbf{p}}_n}{2m_n} + V(\mathbf{r}_1, \mathbf{r}_2 \cdots \mathbf{r}_N, t)$$

$$= -\frac{\hbar^2}{2} \sum_{n=1}^{N} \frac{1}{m_n} \nabla_n^2 + V(\mathbf{r}_1, \mathbf{r}_2 \cdots \mathbf{r}_N, t)$$

However, complications can arise in the many-body problem. Since the potential energy depends on the spatial arrangement of the particles, the kinetic energy will also depend on the spatial configuration to conserve energy. The motion due to any one particle will vary due to the motion of all the other particles in the system. For this reason cross terms for kinetic energy may appear in the Hamiltonian; a mix of the gradients for two particles:

$$-\frac{\hbar^2}{2M} \nabla_i \cdot \nabla_j$$

where M denotes the mass of the collection of particles resulting in this extra kinetic energy. Terms of this form are known as mass polarization terms, and appear in the Hamiltonian of many electron atoms.

For N interacting particles, i.e. particles which interact mutually and constitute a many-body situation, the potential energy function V is not simply a sum of the separate potentials (and certainly not a product, as this is dimensionally incorrect). The potential energy function can only be written as above: a function of all the spatial positions of each particle.

For non-interacting particles, i.e. particles which do not interact mutually and move independently, the potential of the system is the sum of the separate potential energy for each particle, that is:

$$V = \sum_{i=1}^{N} V(\mathbf{r}_i, t) = V(\mathbf{r}_1, t) + V(\mathbf{r}_2, t) + \cdots + V(\mathbf{r}_N, t)$$

The general form of the Hamiltonian in this case is:

$$\begin{aligned}\hat{H} &= -\frac{\hbar^2}{2}\sum_{i=1}^{N}\frac{1}{m_i}\nabla_i^2 + \sum_{i=1}^{N}V_i \\ &= \sum_{i=1}^{N}\left(-\frac{\hbar^2}{2m_i}\nabla_i^2 + V_i\right) \\ &= \sum_{i=1}^{N}\hat{H}_i\end{aligned}$$

where the sum is taken over all particles and their corresponding potentials; the result is that the Hamiltonian of the system is the sum of the separate Hamiltonians for each particle. This is an idealized situation - in practice the particles are almost always influenced by some potential, and there are many-body interactions. One illustrative example of a two-body interaction where this form would not apply is for electrostatic potentials due to charged particles, because they interact with each other by Coulomb interaction (electrostatic force).

Schrödinger Equation

The Hamiltonian generates the time evolution of quantum states. If $|\psi(t)\rangle$ is the state of the system at time t, then:

$$H|\psi(t)\rangle = i\hbar\frac{\partial}{\partial t}|\psi(t)\rangle.$$

This equation is the Schrödinger equation. It takes the same form as the Hamilton–Jacobi equation, which is one of the reasons *H* is also called the Hamiltonian. Given the state at some initial time (t = 0), we can solve it to obtain the state at any subsequent time. In particular, if *H* is independent of time, then:

$$|\psi(t)\rangle = e^{-iHt/\hbar}|\psi(0)\rangle.$$

The exponential operator on the right hand side of the Schrödinger equation is usually defined by the corresponding power series in *H*. One might notice that taking polynomials or power series of unbounded operators that are not defined everywhere may not make mathematical sense.

Rigorously, to take functions of unbounded operators, a functional calculus is required. In the case of the exponential function, the continuous, or just the holomorphic functional calculus suffices. We note again, however, that for common calculations the physicists' formulation is quite sufficient.

By the *-homomorphism property of the functional calculus, the operator:

$$U \quad e^{iHt}$$

is a unitary operator. It is the *time evolution operator*, or *propagator*, of a closed quantum system. If the Hamiltonian is time-independent, $\{U(t)\}$ form a one parameter unitary group (more than a semigroup); this gives rise to the physical principle of detailed balance.

Dirac Formalism

However, in the more general formalism of Dirac, the Hamiltonian is typically implemented as an operator on a Hilbert space in the following way:

The eigenkets (eigenvectors) of *H*, denoted $|a\rangle$, provide an orthonormal basis for the Hilbert space. The spectrum of allowed energy levels of the system is given by the set of eigenvalues, denoted $\{E_a\}$, solving the equation:

$$H|a\rangle = E_a|a\rangle.$$

Since *H* is a Hermitian operator, the energy is always a real number.

From a mathematically rigorous point of view, care must be taken with the above assumptions. Operators on infinite-dimensional Hilbert spaces need not have eigenvalues (the set of eigenvalues does not necessarily coincide with the spectrum of an operator). However, all routine quantum mechanical calculations can be done using the physical formulation.

Expressions for the Hamiltonian

Following are expressions for the Hamiltonian in a number of situations. Typical ways to classify the expressions are the number of particles, number of dimensions, and the nature of the potential energy function - importantly space and time dependence. Masses are denoted by m, and charges by q.

General Forms for one Particle

Free Particle

The particle is not bound by any potential energy, so the potential is zero and this Hamiltonian is the simplest. For one dimension:

$$\hat{H} = -\frac{\hbar^2}{2m}\frac{\partial^2}{\partial x^2}$$

and in higher dimensions:

$$\hat{H} = -\frac{\hbar^2}{2m}\nabla^2$$

Constant-potential Well

For a particle in a region of constant potential $V = V_0$ (no dependence on space or time), in one dimension, the Hamiltonian is:

$$\hat{H} = -\frac{\hbar^2}{2m}\frac{\partial^2}{\partial x^2} + V_0$$

in three dimensions:

$$\hat{H} = -\frac{\hbar^2}{2m}\nabla^2 + V_0$$

This applies to the elementary "particle in a box" problem, and step potentials.

Simple Harmonic Oscillator

For a simple harmonic oscillator in one dimension, the potential varies with position (but not time), according to:

$$V = \frac{k}{2}x^2 = \frac{m\omega^2}{2}x^2$$

where the angular frequency ω, effective spring constant k, and mass m of the oscillator satisfy:

$$\omega^2 = \frac{k}{m}$$

so the Hamiltonian is:

$$\hat{H} = -\frac{\hbar^2}{2m}\frac{\partial^2}{\partial x^2} + \frac{m\omega^2}{2}x^2$$

For three dimensions, this becomes:

$$\hat{H} = -\frac{{}^2}{2}\nabla^2 + \frac{{}^2}{2}r^2$$

where the three-dimensional position vector r using cartesian coordinates is (**r**, y, z), its magnitude is: $r^2 = \mathbf{r}\cdot\mathbf{r} = |\mathbf{r}|^2 = x^2 + y^2 + z^2$.

Writing the Hamiltonian out in full shows it is simply the sum of the one-dimensional Hamiltonians in each direction:

$$\hat{H} = -\frac{\hbar^2}{2m}\left(\frac{\partial^2}{\partial x^2} + \frac{\partial^2}{\partial y^2} + \frac{\partial^2}{\partial z^2}\right) + \frac{m\omega^2}{2}(x^2 + y^2 + z^2)$$

$$=\left(-\frac{\hbar^2}{2m}\frac{\partial^2}{\partial x^2}+\frac{m\omega^2}{2}x^2\right)+\left(-\frac{\hbar^2}{2m}\frac{\partial^2}{\partial y^2}+\frac{m\omega^2}{2}y^2\right)+\left(-\frac{\hbar^2}{2m}\frac{\partial^2}{\partial z^2}+\frac{m\omega^2}{2}z^2\right).$$

Rigid Rotor

For a rigid rotor – i.e. system of particles which can rotate freely about any axes, not bound in any potential (such as free molecules with negligible vibrational degrees of freedom, say due to double or triple chemical bonds), Hamiltonian is:

$$\hat{H}=-\frac{\hbar^2}{2I_{xx}}\hat{J}_x^2-\frac{\hbar^2}{2I_{yy}}\hat{J}_y^2-\frac{\hbar^2}{2I_{zz}}\hat{J}_z^2$$

where I_{xx}, I_{yy}, and I_{yy} are the moment of inertia components (technically the diagonal elements of the moment of inertia tensor), and $\hat{J}_x$, $\hat{J}_y$ and $\hat{J}_z$ are the total angular momentum operators (components), about the x, y, and z axes respectively.

Electrostatic or Coulomb Potential

The Coulomb potential energy for two point charges q_1 and q_2 (i.e. charged particles, since particles have no spatial extent), in three dimensions, is (in SI units - rather than Gaussian units which are frequently used in electromagnetism):

$$V=\frac{q_1q_2}{4\pi\epsilon_0|\mathbf{r}|}$$

However, this is only the potential for one point charge due to another. If there are many charged particles, each charge has a potential energy due to every other point charge (except itself). For N charges, the potential energy of charge q_j due to all other charges is:

$$V_j=\frac{1}{2}\sum_{i\neq j}q_i\phi(\mathbf{r}_i)=\frac{1}{8\pi\varepsilon_0}\sum_{i\neq j}\frac{q_iq_j}{|\mathbf{r}_i-\mathbf{r}_j|}$$

where $\phi(\mathbf{r}_i)$ is the electrostatic potential of charge q_j at $\mathbf{r}_i$. The total potential of the system is then the sum over j:

$$V=\frac{1}{8\pi\varepsilon_0}\sum_{j=1}^{N}\sum_{i\neq j}\frac{q_iq_j}{|\mathbf{r}_i-\mathbf{r}_j|}$$

so the Hamiltonian is:

$$\hat{H}=-\frac{\hbar^2}{2}\sum_{j=1}^{N}\frac{1}{m_j}\nabla_j^2+\frac{1}{8\pi\varepsilon_0}\sum_{j=1}^{N}\sum_{i\neq j}\frac{q_iq_j}{|\mathbf{r}_i-\mathbf{r}_j|}$$

$$=\sum_{j=1}^{N}\left(-\frac{\hbar^2}{2m_j}\nabla_j^2+\frac{1}{8\pi\varepsilon_0}\sum_{i\neq j}\frac{q_iq_j}{|\mathbf{r}_i-\mathbf{r}_j|}\right)$$

Electric Dipole in an Electric Field

For an electric dipole moment **d** constituting charges of magnitude q, in a uniform, electrostatic field (time-independent) **E**, positioned in one place, the potential is:

$$V = -\hat{\mathbf{d}} \cdot \mathbf{E}$$

the dipole moment itself is the operator:

$$\hat{\mathbf{d}} = q\hat{\mathbf{r}}$$

Since the particle is stationary, there is no translational kinetic energy of the dipole, so the Hamiltonian of the dipole is just the potential energy:

$$\hat{H} = -\hat{\mathbf{d}} \cdot \mathbf{E} = -q\hat{\mathbf{r}} \cdot \mathbf{E}$$

Magnetic Dipole in a Magnetic Field

For a magnetic dipole moment μ in a uniform, magnetostatic field (time-independent) **B**, positioned in one place, the potential is:

$$V = -\mu \cdot \mathbf{B}$$

Since the particle is stationary, there is no translational kinetic energy of the dipole, so the Hamiltonian of the dipole is just the potential energy:

$$\hat{H} = -\mu \cdot \mathbf{B}$$

For a Spin-½ particle, the corresponding spin magnetic moment is:

$$\mu_S = \frac{g_s e}{2m}\mathbf{S}$$

where g_s is the spin gyromagnetic ratio (a.k.a. "spin g-factor"), e is the electron charge, **S** is the spin operator vector, whose components are the Pauli matrices, hence,

$$\hat{H} = \frac{g_s e}{2m}\mathbf{S} \cdot \mathbf{B}$$

Charged Particle in an Electromagnetic Field

For a charged particle q in an electromagnetic field, described by the scalar potential ϕ and vector potential **A**, there are two parts to the Hamiltonian to substitute for. The momentum operator $\hat{\Pi}$ must be replaced by the canonical momentum operator, which includes a contribution from the **A** field and fulfils the canonical commutation relation, must be quantized:

$$\hat{\Pi} = \hat{\mathbf{P}} - q\mathbf{A}$$

where $\hat{\mathbf{P}}$ is the kinetic momentum operator given as the usual momentum operator:

$$\hat{\Pi} = -i\hbar\nabla$$

so the corresponding kinetic energy operator is:

$$\hat{T} = \frac{\hat{\mathbf{P}}\cdot\hat{\mathbf{P}}}{2m} = \frac{1}{2m}\left(\hat{\Pi} - q\mathbf{A}\right)^2$$

and the potential energy, which is due to the ϕ field is given by:

$$\hat{V} = q\phi$$

Casting all of these into the Hamiltonian gives:

$$\hat{H} = \frac{1}{2m}\left(-i\hbar\nabla - q\mathbf{A}\right)^2 + q\phi$$

Energy Eigenket Degeneracy, Symmetry and Conservation Laws

In many systems, two or more energy eigenstates have the same energy. A simple example of this is a free particle, whose energy eigenstates have wavefunctions that are propagating plane waves. The energy of each of these plane waves is inversely proportional to the square of its wavelength. A wave propagating in the x direction is a different state from one propagating in the y direction, but if they have the same wavelength, then their energies will be the same. When this happens, the states are said to be *degenerate*.

It turns out that degeneracy occurs whenever a nontrivial unitary operator U commutes with the Hamiltonian. To see this, suppose that $|a\rangle$ is an energy eigenket. Then $U|a\rangle$ is an energy eigenket with the same eigenvalue, since,

$$UH|a\rangle = UE_a|a\rangle = E_a(U|a\rangle) = H(U|a\rangle)$$

Since U is nontrivial, at least one pair of $|a\rangle$ and $U|a\rangle$ must represent distinct states. Therefore, H has at least one pair of degenerate energy eigenkets. In the case of the free particle, the unitary operator which produces the symmetry is the rotation operator, which rotates the wavefunctions by some angle while otherwise preserving their shape.

The existence of a symmetry operator implies the existence of a conserved observable. Let G be the Hermitian generator of U:

$$U = I - i\epsilon G + O(\epsilon^2)$$

It is straightforward to show that if U commutes with H, then so does G:

$$[H, G] = 0$$

Therefore,

$$\frac{\partial}{\partial t}\langle\psi(t)|G|\psi(t)\rangle=\frac{1}{i\hbar}\langle\psi(t)|[G,H]|\psi(t)\rangle=0$$

In obtaining this result, we have used the Schrödinger equation, as well as its dual,

$$\langle\psi(t)|H=-i\hbar\frac{\partial}{\partial t}\langle\psi(t)|.$$

Thus, the expected value of the observable G is conserved for any state of the system. In the case of the free particle, the conserved quantity is the angular momentum.

Hamilton's Equations

Hamilton's equations in classical Hamiltonian mechanics have a direct analogy in quantum mechanics. Suppose we have a set of basis states $\{|n\rangle\}$, which need not necessarily be eigenstates of the energy. For simplicity, we assume that they are discrete, and that they are orthonormal, i.e.,

$$\langle n'|n\rangle=\delta_{nn'}$$

Note that these basis states are assumed to be independent of time. We will assume that the Hamiltonian is also independent of time.

The instantaneous state of the system at time t, $|\psi(t)\rangle$ can be expanded in terms of these basis states:

$$|\psi(t)\rangle=\sum_n a_n(t)|n\rangle$$

where,

$$a_n(t)=\langle n|\psi(t)\rangle$$

The coefficients $a_n(t)$ are complex variables. We can treat them as coordinates which specify the state of the system, like the position and momentum coordinates which specify a classical system. Like classical coordinates, they are generally not constant in time, and their time dependence gives rise to the time dependence of the system as a whole.

The expectation value of the Hamiltonian of this state, which is also the mean energy, is:

$$\langle H(t)\rangle\overset{\text{def}}{=}\langle\psi(t)|H|\psi(t)\rangle=\sum_{nn'}a_{n'}^{*}a_n\langle n'|H|n\rangle$$

where the last step was obtained by expanding $|\psi(t)\rangle$ in terms of the basis states.

Each $a_n(t)$ actually corresponds to two independent degrees of freedom, since the variable has a real part and an imaginary part. We now perform the following trick: instead of using the real and imaginary parts as the independent variables, we use $a_n(t)$ and its complex conjugate $a_n^*(t)$. With this choice of independent variables, we can calculate the partial derivative:

$$\frac{\partial \langle H \rangle}{\partial a_{n'}^*} = \sum_n a_n \langle n' | H | n \rangle = \langle n' | H | \psi \rangle$$

By applying Schrödinger's equation and using the orthonormality of the basis states, this further reduces to,

$$\frac{\partial \langle H \rangle}{\partial a_{n'}^*} = i\hbar \frac{\partial a_{n'}}{\partial t}$$

Similarly, one can show that,

$$\frac{\partial \langle H \rangle}{\partial a_n} = -i\hbar \frac{\partial a_n^*}{\partial t}$$

If we define "conjugate momentum" variables π_n by:

$$\pi_n(t) = i\hbar a_n^*(t)$$

then the above equations become,

$$\frac{\partial \langle H \rangle}{\partial \pi_n} = \frac{\partial a_n}{\partial t}, \quad \frac{\partial \langle H \rangle}{\partial a_n} = -\frac{\partial \pi_n}{\partial t}$$

which is precisely the form of Hamilton's equations, with the s as the generalized coordinates, the π_n s as the conjugate momenta, and $\langle H \rangle$ taking the place of the classical Hamiltonian.

MATHEMATICAL FORMULATIONS

The mathematical formulations of quantum mechanics are those mathematical formalisms that permit a rigorous description of quantum mechanics. Such are distinguished from mathematical formalisms for theories developed prior to the early 1900s by the use of abstract mathematical structures, such as infinite-dimensional Hilbert spaces and operators on these spaces. Many of these structures are drawn from functional analysis, a research area within pure mathematics that was influenced in part by the needs of quantum mechanics. In brief, values of physical observables such as energy and momentum were no longer considered as values of functions on phase space, but as eigenvalues; more precisely as spectral values of linear operators in Hilbert space.

These formulations of quantum mechanics continue to be used today. At the heart of the description are ideas of *quantum state* and *quantum observables* which are radically different from those used in previous models of physical reality. While the mathematics permits calculation of many quantities that can be measured experimentally, there is a definite theoretical limit to values that can be simultaneously measured. This limitation was first elucidated by Heisenberg through a thought experiment, and is represented mathematically in the new formalism by the non-commutativity of operators representing quantum observables.

Prior to the emergence of quantum mechanics as a separate theory, the mathematics used in physics consisted mainly of formal mathematical analysis, beginning with calculus, and increasing in complexity up to differential geometry and partial differential equations. Probability theory was used in statistical mechanics. Geometric intuition played a strong role in the first two and, accordingly, theories of relativity were formulated entirely in terms of geometric concepts. The phenomenology of quantum physics arose roughly between 1895 and 1915, and for the 10 to 15 years before the emergence of quantum theory physicists continued to think of quantum theory within the confines of what is now called classical physics, and in particular within the same mathematical structures. The most sophisticated example of this is the Sommerfeld–Wilson–Ishiwara quantization rule, which was formulated entirely on the classical phase space.

"Old Quantum Theory" and the Need for New Mathematics

In the 1890s, Planck was able to derive the blackbody spectrum which was later used to avoid the classical ultraviolet catastrophe by making the unorthodox assumption that, in the interaction of electromagnetic radiation with matter, energy could only be exchanged in discrete units which he called quanta. Planck postulated a direct proportionality between the frequency of radiation and the quantum of energy at that frequency. The proportionality constant, h, is now called Planck's constant in his honor.

In 1905, Einstein explained certain features of the photoelectric effect by assuming that Planck's energy quanta were actual particles, which were later dubbed photons.

All of these developments were phenomenological and challenged the theoretical physics of the time. Bohr and Sommerfeld went on to modify classical mechanics in an attempt to deduce the Bohr model from first principles. They proposed that, of all closed classical orbits traced by a mechanical system in its phase space, only the ones that enclosed an area which was a multiple of Planck's constant were actually allowed. The most sophisticated version of this formalism was the so-called Sommerfeld–Wilson–Ishiwara quantization. Although the Bohr model of the hydrogen atom could be explained in this way, the spectrum of the helium atom (classically an unsolvable 3-body problem) could not be predicted. The mathematical status of quantum theory remained uncertain for some time.

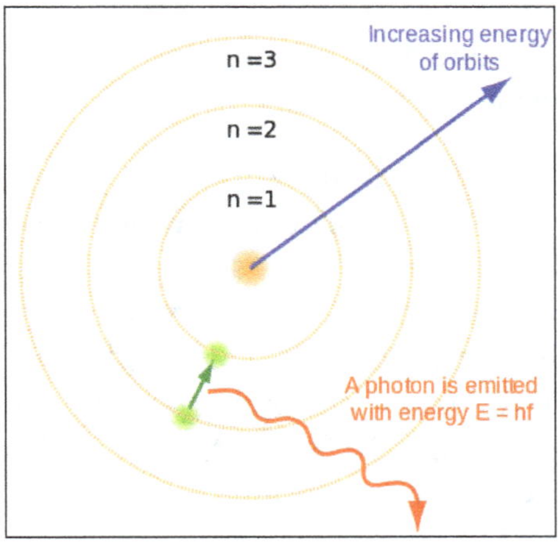

In 1923 de Broglie proposed that wave–particle duality applied not only to photons but to electrons and every other physical system.

The situation changed rapidly in the years 1925–1930, when working mathematical foundations were found through the groundbreaking work of Erwin Schrödinger, Werner Heisenberg, Max Born, Pascual Jordan, and the foundational work of John von Neumann, Hermann Weyl and Paul Dirac, and it became possible to unify several different approaches in terms of a fresh set of ideas. The physical interpretation of the theory was also clarified in these years after Werner Heisenberg discovered the uncertainty relations and Niels Bohr introduced the idea of complementarity.

The "New Quantum Theory"

Werner Heisenberg's matrix mechanics was the first successful attempt at replicating the observed quantization of atomic spectra. Later in the same year, Schrödinger created his wave mechanics. Schrödinger's formalism was considered easier to understand, visualize and calculate as it led to differential equations, which physicists were already familiar with solving. Within a year, it was shown that the two theories were equivalent.

Schrödinger himself initially did not understand the fundamental probabilistic nature of quantum mechanics, as he thought that the absolute square of the wave function of an electron should be interpreted as the charge density of an object smeared out over an extended, possibly infinite, volume of space. It was Max Born who introduced the interpretation of the absolute square of the wave function as the probability distribution of the position of a *pointlike* object. Born's idea was soon taken over by Niels Bohr in Copenhagen who then became the "father" of the Copenhagen interpretation of quantum mechanics. Schrödinger's wave function can be seen to be closely related to the classical Hamilton–Jacobi equation. The correspondence to classical mechanics was even more explicit, although somewhat more formal, in Heisenberg's matrix mechanics. In his PhD thesis project, Paul Dirac discovered that the equation for the operators in the Heisenberg representation, as it is now called, closely translates to classical equations for the dynamics of certain quantities in the Hamiltonian formalism of classical mechanics, when one expresses them through Poisson brackets, a procedure now known as canonical quantization.

To be more precise, already before Schrödinger, the young postdoctoral fellow Werner Heisenberg invented his matrix mechanics, which was the first correct quantum mechanics – the essential breakthrough. Heisenberg's matrix mechanics formulation was based on algebras of infinite matrices, a very radical formulation in light of the mathematics of classical physics, although he started from the index-terminology of the experimentalists of that time, not even aware that his "index-schemes" were matrices, as Born soon pointed out to him. In fact, in these early years, linear algebra was not generally popular with physicists in its present form.

Although Schrödinger himself after a year proved the equivalence of his wave-mechanics and Heisenberg's matrix mechanics, the reconciliation of the two approaches and their modern abstraction as motions in Hilbert space is generally attributed to Paul Dirac, who wrote a lucid account in his 1930 classic *The Principles of Quantum Mechanics*. He is the third, and possibly most important, pillar of that field (he soon was the only one to have discovered a relativistic generalization of the theory). In his above-mentioned account, he introduced the bra–ket notation, together with an abstract formulation in terms of the Hilbert space used in functional analysis; he showed that Schrödinger's and Heisenberg's approaches were two different representations of the same theory, and found a third, most general one, which represented the dynamics of the system. His work was particularly fruitful in all kinds of generalizations of the field.

The first complete mathematical formulation of this approach, known as the Dirac–von Neumann axioms, is generally credited to John von Neumann's 1932 book *Mathematical Foundations of Quantum Mechanics*, although Hermann Weyl had already referred to Hilbert spaces (which he called *unitary spaces*) in his 1927 classic paper and book. It was developed in parallel with a new approach to the mathematical spectral theory based on linear operators rather than the quadratic forms that were David Hilbert's approach a generation earlier. Though theories of quantum mechanics continue to evolve to this day, there is a basic framework for the mathematical formulation of quantum mechanics which underlies most approaches and can be traced back to the mathematical work of John von Neumann. In other words, discussions about *interpretation* of the theory, and extensions to it, are now mostly conducted on the basis of shared assumptions about the mathematical foundations.

Later Developments

The application of the new quantum theory to electromagnetism resulted in quantum field theory, which was developed starting around 1930. Quantum field theory has driven the development of more sophisticated formulations of quantum mechanics, of which the ones presented here are simple special cases:

- Path integral formulation.
- Phase space formulation of quantum mechanics & geometric quantization.
- Signed particle formulation.
- Quantum field theory in curved spacetime.
- Axiomatic, algebraic and constructive quantum field theory.
- C* algebra formalism.
- Generalized statistical model of quantum mechanics.

On a different front, von Neumann originally dispatched quantum measurement with his infamous postulate on the collapse of the wavefunction, raising a host of philosophical problems. Over the intervening 70 years, the *problem of measurement* became an active research area and itself spawned some new formulations of quantum mechanics:

- Relative state/Many-worlds interpretation of quantum mechanics.
- Decoherence.
- Consistent histories formulation of quantum mechanics.
- Quantum logic formulation of quantum mechanics.

A related topic is the relationship to classical mechanics. Any new physical theory is supposed to reduce to successful old theories in some approximation. For quantum mechanics, this translates into the need to study the so-called classical limit of quantum mechanics. Also, as Bohr emphasized, human cognitive abilities and language are inextricably linked to the classical realm, and so classical descriptions are intuitively more accessible than quantum ones. In particular, quantization, namely the construction of a quantum theory whose classical limit is a given and known classical theory, becomes an important area of quantum physics in itself.

Finally, some of the originators of quantum theory (notably Einstein and Schrödinger) were unhappy with what they thought were the philosophical implications of quantum mechanics. In particular, Einstein took the position that quantum mechanics must be incomplete, which motivated research into so-called hidden-variable theories. The issue of hidden variables has become in part an experimental issue with the help of quantum optics:

- De Broglie–Bohm–Bell pilot wave formulation of quantum mechanics.

- Bell's inequalities.
- Kochen–Specker theorem.

Mathematical Structure of Quantum Mechanics

A physical system is generally described by three basic ingredients: states; observables; and dynamics (or law of time evolution) or, more generally, a group of physical symmetries. A classical description can be given in a fairly direct way by a phase space model of mechanics: states are points in a symplectic phase space, observables are real-valued functions on it, time evolution is given by a one-parameter group of symplectic transformations of the phase space, and physical symmetries are realized by symplectic transformations. A quantum description normally consists of a Hilbert space of states, observables are self adjoint operators on the space of states, time evolution is given by a one-parameter group of unitary transformations on the Hilbert space of states, and physical symmetries are realized by unitary transformations. (It is possible, to map this Hilbert-space picture to a phase space formulation, invertibly).

Postulates of Quantum Mechanics

The following summary of the mathematical framework of quantum mechanics can be partly traced back to the Dirac–von Neumann axioms.

- Each physical system is associated with a (topologically) separable complex Hilbert space H with inner product $\langle\varphi|\psi\rangle$. Rays (that is, subspaces of *complex* dimension 1) in H are associated with quantum states of the system. In other words, quantum states can be identified with equivalence classes of vectors of length 1 in H, where two vectors represent the same state if they differ only by a phase factor. *Separability* is a mathematically convenient hypothesis, with the physical interpretation that countably many observations are enough to uniquely determine the state. "A quantum mechanical state is a *ray* in projective Hilbert space, not a *vector*. Many textbooks fail to make this distinction, which could be partly a result of the fact that the Schrödinger equation itself involves Hilbert-space "vectors", with the result that the imprecise use of "state vector" rather than *ray* is very difficult to avoid."
- The Hilbert space of a composite system is the Hilbert space tensor product of the state spaces associated with the component systems. For a non-relativistic system consisting of a finite number of distinguishable particles, the component systems are the individual particles.
- Physical symmetries act on the Hilbert space of quantum states unitarily or antiunitarily due to Wigner's theorem (supersymmetry is another matter entirely).

- Physical observables are represented by Hermitian matrices on H.
- The expectation value (in the sense of probability theory) of the observable A for the system in state represented by the unit vector $\psi \in H$ is:

  $$\langle \psi \mid A \mid \psi \rangle$$

- By spectral theory, we can associate a probability measure to the values of A in any state ψ. We can also show that the possible values of the observable A in any state must belong to the spectrum of A. In the special case A has only discrete spectrum, the possible outcomes of measuring A are its eigenvalues. More precisely, if we represent the state ψ in the basis formed by the eigenvectors of A, then the square of the modulus of the component attached to a given eigenvector is the probability of observing its corresponding eigenvalue.
- More generally, a state can be represented by a so-called density operator, which is a trace class, nonnegative self-adjoint operator ρ normalized to be of trace 1. The expected value of A in the state ρ is:

  $$\operatorname{tr}(A\rho)$$

- If ρ_ψ is the orthogonal projector onto the one-dimensional subspace of H spanned by $|\psi\rangle$, then:

  $$\operatorname{tr}(A\rho_\psi) = \langle \psi \mid A \mid \psi \rangle$$

- Density operators are those that are in the closure of the convex hull of the one-dimensional orthogonal projectors. Conversely, one-dimensional orthogonal projectors are extreme points of the set of density operators. Physicists also call one-dimensional orthogonal projectors *pure states* and other density operators *mixed states*.

One can in this formalism state Heisenberg's uncertainty principle and prove it as a theorem, although the exact historical sequence of events, concerning who derived what and under which framework, is the subject of historical investigations.

Furthermore, to the postulates of quantum mechanics one should also add basic statements on the properties of spin and Pauli's exclusion principle.

Pictures of Dynamics

In the so-called Schrödinger picture of quantum mechanics, the dynamics is given as follows:

The time evolution of the state is given by a differentiable function from the real numbers **R**, representing instants of time, to the Hilbert space of system states. This map is characterized by a differential equation as follows: If $|\psi(t)\rangle$ denotes the state of the system at any one time *t*, the following Schrödinger equation holds:

$$i\hbar\frac{d}{dt}|\psi(t)\rangle = H|\psi(t)\rangle$$

where *H* is a densely defined self-adjoint operator, called the system Hamiltonian, *i* is the imaginary unit and $\hbar$ is the reduced Planck constant. As an observable, *H* corresponds to the total energy of the system.

Alternatively, by Stone's theorem one can state that there is a strongly continuous one-parameter unitary map U(t): $H \rightarrow H$ such that:

$$|\psi(t+s)\rangle = U(t)|\psi(s)\rangle$$

for all times s, t. The existence of a self-adjoint Hamiltonian *H* such that:

$$U(t) = e^{-(i/\hbar)tH}$$

is a consequence of Stone's theorem on one-parameter unitary groups. It is assumed that *H* does not depend on time and that the perturbation starts at $t_0 = 0$; otherwise one must use the Dyson series, formally written as:

$$U(t) = \mathcal{T}\left[\exp\left(-\frac{i}{\hbar}\int_{t_0}^{t} \mathrm{d}t'\, H(t')\right)\right],$$

where $\mathcal{T}$ is Dyson's time-ordering symbol.

This symbol permutes a product of noncommuting operators of the form:

$$B_1(t_1)\cdot B_2(t_2)\cdot\ldots\cdot B_n(t_n)$$

into the uniquely determined re-ordered expression:

$$B_{i_1}(t_{i_1})\cdot B_{i_2}(t_{i_2})\cdot\ldots\cdot B_{i_n}(t_{i_n}) \text{ With } t_{i_1} \geq t_{i_2} \geq \ldots \geq t_{i_n}.$$

The result is a causal chain, the primary cause in the past on the utmost r.h.s., and finally the present effect on the utmost l.h.s.

The Heisenberg picture of quantum mechanics focuses on observables and instead of considering states as varying in time, it regards the states as fixed and the observables as changing. To go from the Schrödinger to the Heisenberg picture one needs to define time-independent states and time-dependent operators thus:

$$|\psi\rangle = |\psi(0)\rangle$$
$$A(t) = U(-t)AU(t).$$

It is then easily checked that the expected values of all observables are the same in both pictures:

$$\langle\psi \mid A(t) \mid \psi\rangle = \langle\psi(t) \mid A \mid \psi(t)\rangle$$

and that the time-dependent Heisenberg operators satisfy:

$$\frac{d}{dt}A(t) = \frac{i}{\hbar}[H, A(t)] + \frac{\partial A(t)}{\partial t},$$

which is true for time-dependent $A = A(t)$. Notice the commutator expression is purely formal when one of the operators is unbounded. One would specify a representation for the expression to make sense of it.

The so-called Dirac picture or interaction picture has time-dependent *states* and observables, evolving with respect to different Hamiltonians. This picture is most useful when the evolution of the observables can be solved exactly, confining any complications to the evolution of the states. For this reason, the Hamiltonian for the observables is called "free Hamiltonian" and the Hamiltonian for the states is called "interaction Hamiltonian". In symbols:

$$i\hbar\frac{d}{dt}\left|\psi(t)\right\rangle = H_{\text{int}}(t)\left|\psi(t)\right\rangle$$

$$i\hbar\frac{d}{dt}A(t) = [A(t), H_0].$$

The interaction picture does not always exist, though. In interacting quantum field theories, Haag's theorem states that the interaction picture does not exist. This is because the Hamiltonian cannot be split into a free and an interacting part within a superselection sector. Moreover, even if in the Schrödinger picture the Hamiltonian does not depend on time, e.g. $H = H_0 + V$, in the interaction picture it does, at least, if V does not commute with H_0, since:

$$H_{\text{int}}(t) \equiv e^{(i/\hbar)tH_0}\, V\, e^{(-i/\hbar)tH_0}.$$

So the above-mentioned Dyson-series has to be used anyhow.

The Heisenberg picture is the closest to classical Hamiltonian mechanics (for example, the commutators appearing in the above equations directly translate into the classical Poisson brackets); but this is already rather "high-browed", and the Schrödinger picture is considered easiest to visualize and understand by most people, to judge from pedagogical accounts of quantum mechanics. The Dirac picture is the one used in perturbation theory, and is specially associated to quantum field theory and many-body physics.

Similar equations can be written for any one-parameter unitary group of symmetries of the physical system. Time would be replaced by a suitable coordinate parameterizing the unitary group (for instance, a rotation angle, or a translation distance) and the Hamiltonian would be replaced by the conserved quantity associated to the symmetry (for instance, angular or linear momentum).

Representations of Heisenberg's Canonical Commutation Relations

The original form of the Schrödinger equation depends on choosing a particular representation of Heisenberg's canonical commutation relations. The Stone–von Neumann theorem dictates that all irreducible representations of the finite-dimensional Heisenberg commutation relations are unitarily equivalent. A systematic understanding of its consequences has led to the phase space formulation of quantum mechanics, which works in full phase space instead of Hilbert space, so then with a more intuitive link to the classical limit thereof. This picture also simplifies considerations of quantization, the deformation extension from classical to quantum mechanics.

The quantum harmonic oscillator is an exactly solvable system where the different representations are easily compared. There, apart from the Heisenberg, or Schrödinger (position or momentum), or phase-space representations, one also encounters the Fock (number) representation and the Segal–Bargmann (Fock-space or coherent state) representation (named after Irving Segal and Valentine Bargmann). All four are unitarily equivalent.

Time as an Operator

The framework presented so far singles out time as *the* parameter that everything depends on. It is possible to formulate mechanics in such a way that time becomes itself an observable associated to a self-adjoint operator. At the classical level, it is possible to arbitrarily parameterize the trajectories of particles in terms of an unphysical parameter s, and in that case the time t becomes an additional generalized coordinate of the physical system. At the quantum level, translations in s would be generated by a "Hamiltonian" $H - E$, where E is the energy operator and H is the "ordinary" Hamiltonian. However, since s is an unphysical parameter, *physical* states must be left invariant by "s-evolution", and so the physical state space is the kernel of $H - E$ (this requires the use of a rigged Hilbert space and a renormalization of the norm).

This is related to the quantization of constrained systems and quantization of gauge theories. It is also possible to formulate a quantum theory of "events" where time becomes an observable.

Spin

In addition to their other properties, all particles possess a quantity called spin, an *intrinsic angular momentum*. Despite the name, particles do not literally spin around an axis, and quantum mechanical spin has no correspondence in classical physics. In the position representation, a spinless wavefunction has position **r** and time t as continuous variables, $\psi = \psi(\mathbf{r}, t)$, for spin wavefunctions the spin is an additional discrete variable: $\psi = \psi(\mathbf{r}, t, \sigma)$, where σ takes the values:

$$\sigma = -S\hbar, -(S-1)\hbar, \ldots, 0, \ldots, +(S-1)\hbar, +S\hbar.$$

That is, the state of a single particle with spin S is represented by a $(2S + 1)$-component spinor of complex-valued wave functions.

Two classes of particles with *very different* behaviour are bosons which have integer spin (S = 0, 1, 2...), and fermions possessing half-integer spin (S = ½, ³⁄₂, ⁵⁄₂, ...).

Pauli's Principle

The property of spin relates to another basic property concerning systems of N identical particles: Pauli's exclusion principle, which is a consequence of the following permutation behaviour of an N-particle wave function; again in the position representation one must postulate that for the transposition of any two of the N particles one always should have:

$$\psi(\ldots, \mathbf{r}_i, \sigma_i, \ldots, \mathbf{r}_j, \sigma_j, \ldots) = (-1)^{2S} \cdot \psi(\ldots, \mathbf{r}_j, \sigma_j, \ldots, \mathbf{r}_i, \sigma_i, \ldots)$$

i.e., on transposition of the arguments of any two particles the wavefunction should *reproduce*, apart from a prefactor $(-1)^{2S}$ which is +1 for bosons, but (−1) for fermions. Electrons are fermions with S = 1/2; quanta of light are bosons with S = 1. In nonrelativistic quantum mechanics all particles are either bosons or fermions; in relativistic quantum theories also "supersymmetric" theories exist, where a particle is a linear combination of a bosonic and a fermionic part. Only in dimension d = 2 can one construct entities where $(-1)^{2S}$ is replaced by an arbitrary complex number with magnitude 1, called anyons.

Although *spin* and the *Pauli principle* can only be derived from relativistic generalizations of quantum mechanics the properties mentioned in the last two paragraphs belong to the basic postulates already in the non-relativistic limit. Especially, many important properties in natural science, e.g. the periodic system of chemistry, are consequences of the two properties.

Problem of Measurement

The picture given in the preceding paragraphs is sufficient for description of a completely isolated system. However, it fails to account for one of the main differences between quantum mechanics and classical mechanics, that is, the effects of measurement. The von Neumann description of quantum measurement of an observable A, when the system is prepared in a pure state ψ is the following (note, however, that von Neumann's description dates back to the 1930s and is based on experiments as performed during that time – more specifically the Compton–Simon experiment; it is not applicable to most present-day measurements within the quantum domain):

- Let A have spectral resolution:

  $$A = \int \lambda \, d\, \mathrm{E}_A(\lambda),$$

  where E_A is the resolution of the identity (also called projection-valued measure) associated to A. Then the probability of the measurement outcome lying in an interval B of $\mathbf{R}$ is $|\mathrm{E}_A(B)\, \psi|^2$. In other words, the probability is obtained by integrating the characteristic function of B against the countably additive measure:

  $$\langle \psi \,|\, \mathrm{E}_A \psi \rangle$$

- If the measured value is contained in B, then immediately after the measurement, the system will be in the (generally non-normalized) state $\mathrm{E}_A(B)\psi$. If the measured value does not lie in B, replace B by its complement for the above state.

For example, suppose the state space is the n-dimensional complex Hilbert space C^n and A is a Hermitian matrix with eigenvalues λ_i, with corresponding eigenvectors ψ_i. The projection-valued measure associated with A, E_A, is then:

$$\mathrm{E}_A(B) = |\psi_i\rangle\langle\psi_i|,$$

where B is a Borel set containing only the single eigenvalue λ_i. If the system is prepared in state:

$$|\psi\rangle$$

Then the probability of a measurement returning the value λ_i can be calculated by integrating the spectral measure:

$$\langle \psi \,|\, \mathrm{E}_A \psi \rangle$$

over B_i. This gives trivially:

$$\langle \psi \,|\, \psi_i \rangle\langle \psi_i \,|\, \psi \rangle = |\langle \psi \,|\, \psi_i \rangle|$$

The characteristic property of the von Neumann measurement scheme is that repeating the same measurement will give the same results. This is also called the *projection postulate.*

A more general formulation replaces the projection-valued measure with a positive-operator valued measure (POVM). To illustrate, take again the finite-dimensional case. Here we would replace the rank-1 projections:

$$|\psi_i\rangle\langle\psi_i|$$

by a finite set of positive operators:

$$F_i F_i^*$$

whose sum is still the identity operator as before (the resolution of identity). Just as a set of possible outcomes $\{\lambda_1 \ldots \lambda_n\}$ is associated to a projection-valued measure, the same can be said for a POVM. Suppose the measurement outcome is λ_i. Instead of collapsing to the (unnormalized) state:

$$|\psi_i\rangle\langle\psi_i|\psi\rangle$$

after the measurement, the system now will be in the state:

$$F_i|\psi\rangle$$

Since the $F_i F_i^*$ operators need not be mutually orthogonal projections, the projection postulate of von Neumann no longer holds.

The same formulation applies to general mixed states.

In von Neumann's approach, the state transformation due to measurement is distinct from that due to time evolution in several ways. For example, time evolution is deterministic and unitary whereas measurement is non-deterministic and non-unitary. However, since both types of state transformation take one quantum state to another, this difference was viewed by many as unsatisfactory. The POVM formalism views measurement as one among many other quantum operations, which are described by completely positive maps which do not increase the trace.

In any case it seems that the above-mentioned problems can only be resolved if the time evolution included not only the quantum system, but also, and essentially, the classical measurement apparatus.

SCHRÖDINGER EQUATION

The Schrödinger equation is a linear partial differential equation that describes the wave function or state function of a quantum-mechanical system. It is a key result in quantum mechanics, and its discovery was a significant landmark in the development of the subject. The equation is named after Erwin Schrödinger, who postulated the equation in 1925, and published it in 1926, forming the basis for the work that resulted in his Nobel Prize in Physics in 1933.

In classical mechanics, Newton's second law (**F** = m**a**) is used to make a mathematical prediction as to what path a given physical system will take over time following a set of known initial conditions. Solving this equation gives the position and the momentum of the physical system as a function of the external force **F** on the system. Those two parameters are sufficient to describe its state at each time instant. In quantum mechanics, the analogue of Newton's law is Schrödinger's equation.

The concept of a wave function is a fundamental postulate of quantum mechanics; the wave function defines the state of the system at each spatial position, and time. Using these postulates, Schrödinger's equation can be derived from the fact that the time-evolution operator must be unitary, and must therefore be generated by the exponential of a self-adjoint operator, which is the quantum Hamiltonian.

In the Copenhagen interpretation of quantum mechanics, the wave function is the most complete description that can be given of a physical system. Solutions to Schrödinger's equation describe not only molecular, atomic, and subatomic systems, but also macroscopic systems, possibly even the whole universe. Schrödinger's equation is central to all applications of quantum mechanics, including quantum field theory, which combines special relativity with quantum mechanics. Theories of quantum gravity, such as string theory, also do not modify Schrödinger's equation.

The Schrödinger equation is not the only way to study quantum mechanical systems and make predictions. The other formulations of quantum mechanics include matrix mechanics, introduced by Werner Heisenberg, and the path integral formulation, developed chiefly by Richard Feynman. Paul Dirac incorporated matrix mechanics and the Schrödinger equation into a single formulation.

Time-dependent Equation

The form of the Schrödinger equation depends on the physical situation. The most general form is the time-dependent Schrödinger equation (TDSE), which gives a description of a system evolving with time:

$$i\hbar\frac{d}{dt}|\Psi(t)\rangle = \hat{H}|\Psi(t)\rangle$$

where i is the imaginary unit, $\hbar = \frac{h}{2\pi}$ is the reduced Planck constant, Ψ is the state vector of the quantum system, t is time, and $\hat{H}$ is the Hamiltonian operator. The position-space wave function of the quantum system is nothing but the components in the expansion of the state vector in terms of the position eigenvector $|\mathbf{r}\rangle$. It is a scalar function, expressed as $\Psi(\mathbf{r},t) = \langle\mathbf{r}|\Psi\rangle$. Similarly, the momentum-space wave function can be defined as $\tilde{\Psi}(\mathbf{p},t) = \langle\mathbf{p}|\Psi\rangle$, where $|\mathbf{p}\rangle$ is the momentum eigenvector.

The most famous example is the nonrelativistic Schrödinger equation for the wave function in position space $\Psi(\mathbf{r},t)$ of a single particle subject to a potential $V(\mathbf{r},t)$, such as that due to an electric field:

$$i\hbar\frac{\partial}{\partial t}\Psi(\mathbf{r},t)=\left[\frac{-\hbar^2}{2m}\nabla^2+V(\mathbf{r},t)\right]\Psi(\mathbf{r},t)$$

where m is the particle's mass, and ∇^2 is the Laplacian.

This is also a diffusion equation, but unlike the heat equation, this one is also a wave equation given the imaginary unit present in the transient term.

The term "Schrödinger equation" can refer to both the general equation, or the specific nonrelativistic version. The general equation is indeed quite general, used throughout quantum mechanics, for everything from the Dirac equation to quantum field theory, by plugging in diverse expressions for the Hamiltonian. The specific nonrelativistic version is a strictly classical approximation to reality and yields accurate results in many situations, but only to a certain extent.

To apply the Schrödinger equation, write down the Hamiltonian for the system, accounting for the kinetic and potential energies of the particles constituting the system, then insert it into the Schrödinger equation. The resulting partial differential equation is solved for the wave function, which contains information about the system.

Time-independent Equation

The time-dependent Schrödinger equation described above predicts that wave functions can form standing waves, called stationary states. These states are particularly important as their individual study later simplifies the task of solving the time-dependent Schrödinger equation for any state. Stationary states can also be described by a simpler form of the Schrödinger equation, the time-independent Schrödinger equation (TISE):

$$\hat{H}\,|\Psi\rangle=E\,|\Psi\rangle$$

where E is a constant equal to the energy level of the system. This is only used when the Hamiltonian itself is not dependent on time explicitly. However, even in this case the total wave function still has a time dependency.

In the language of linear algebra, this equation is an eigenvalue equation. Therefore, the wave function is an eigenfunction of the Hamiltonian operator with corresponding eigenvalues E.

As before, the most common manifestation is the nonrelativistic Schrödinger equation for a single particle moving in an electric field (but not a magnetic field):

$$\left[\frac{-\hbar^2}{2m}\nabla^2+V(\mathbf{r})\right]\Psi(\mathbf{r})=E\Psi(\mathbf{r})$$

with definitions as above. Here, the form of the Hamiltonian operator comes from classical mechanics, where the Hamiltonian function is the sum of the kinetic and potential energies. That is, $H = T + V = \frac{\|\mathbf{p}\|^2}{2m} + V(x, y, z)$ for a single particle in the non-relativistic limit.

Derivation

By assuming the Dirac-von Neumann axioms, a modern approach to understanding the Schrödinger equation is outlined as follows. Suppose the wave function $\psi(t_0)$ represents a unit vector defined on a complex Hilbert space at some initial time t_0. The unitarity principle requires that there must exist a linear operator, $\hat{U}(t)$, such that for any time $t > t_0$,

$$\psi(t) = \hat{U}(t)\psi(t_0).$$

Given that: $\psi(t)$ must remain a unit vector, the operator $\hat{U}(t)$ must therefore be a unitary transformation. As such, there exists an exponential map such that $\hat{U}(t) = e^{-i\widehat{\mathcal{H}}t}$ where $\widehat{\mathcal{H}}$ is a Hermitian operator (given by the fact that the Lie algebra of the unitary group is generated by Hermitian operators and if $\widehat{\mathcal{H}}$ is Hermitian, then $-i\hat{\mathcal{H}}$ is skew-Hermitian). Therefore, the first-order Taylor expansion of $\hat{U}(t)$ centered at t_0 takes the form:

$$\hat{U}(t) \approx 1 - i(t - t_0)\widehat{\mathcal{H}}.$$

And so, substituting the above expansion into $\psi(t) = \hat{U}(t)\psi(t_0)$ thus yields:

$$\psi(t) = \psi(t_0) - i(t - t_0)\widehat{\mathcal{H}}\psi(t_0),$$

which, when rearranged and taken in the limit $t \to t_0$, provides an equation of the same form as the Schrödinger equation:

$$i\frac{d\psi}{dt} = \widehat{\mathcal{H}}\psi,$$

where the ordinary definition for the derivative was used. However, the operator $\widehat{\mathcal{H}}$ used here denotes an arbitrary Hermitian operator. Nonetheless, by using the correspondence principle it is possible to show that, in the classical limit and using appropriate units, the expectation value of $\widehat{\mathcal{H}}$ indeed corresponds to the Hamiltonian of the system.

Implications

Energy

The Hamiltonian is constructed in the same manner as in classical mechanics. However, in classical mechanics, the Hamiltonian is a scalar-valued function, whereas in quantum mechanics, it is an operator on a space of functions. It is not surprising that the eigenvalues of $\widehat{H}$ are the energy levels of the system.

Quantization

The Schrödinger equation predicts that if certain properties of a system are measured, the result may be quantized, meaning that only specific discrete values can occur. One example is energy quantization: the energy of an electron in an atom is always one of the quantized energy levels, a fact discovered via atomic spectroscopy. Another example is quantization of angular momentum. This was an assumption in the earlier Bohr model of the atom, but it is a prediction of the Schrödinger equation.

Another result of the Schrödinger equation is that not every measurement gives a quantized result in quantum mechanics. For example, position, momentum, time, and (in some situations) energy can have any value across a continuous range.

Quantum Tunneling

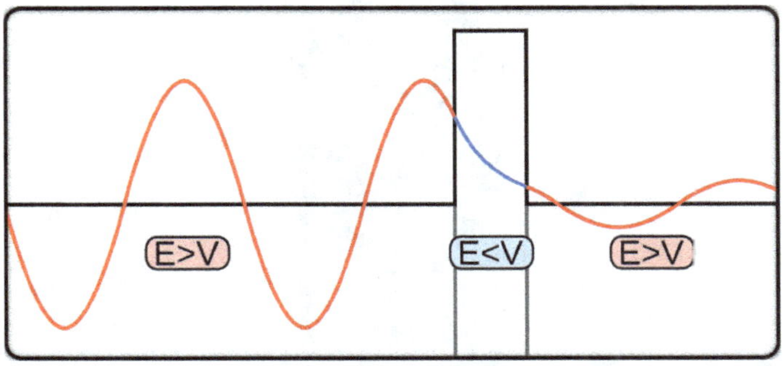

Quantum tunneling through a barrier. A particle coming from the left does not have enough energy to climb the barrier. However, it can sometimes "tunnel" to the other side.

In classical physics, when a ball is rolled slowly up a large hill, it will come to a stop and roll back, because it doesn't have enough energy to get over the top of the hill to the other side. However, the Schrödinger equation predicts that there is a small probability that the ball will get to the other side of the hill, even if it has too little energy to reach the top. This is called quantum tunneling. It is related to the distribution of energy: although the ball's assumed position seems to be on one side of the hill, there is a chance of finding it on the other side.

Particles as Waves

The nonrelativistic Schrödinger equation is a type of partial differential equation called a wave equation. Therefore, it is often said particles can exhibit behavior usually attributed to waves. In some modern interpretations this description is reversed – the quantum state, i.e. wave, is the only genuine physical reality, and under the appropriate conditions it can show features of particle-like behavior. However, Ballentine shows that such an interpretation has problems. Ballentine points out that whilst it is arguable to associate a physical wave with a single particle, there is still only one Schrödinger wave equation for many particles. He points out:

> "If a physical wave field were associated with a particle, or if a particle were identified with a wave packet, then corresponding to N interacting particles there should be N interacting waves in ordinary three-dimensional space. There is one "wave" function in an abstract 3N-dimensional configuration space. The misinterpretation of psi as a physical wave in ordinary space is possible only because the most common applications of quantum mechanics are to one-particle states, for which configuration space and ordinary space are isomorphic."

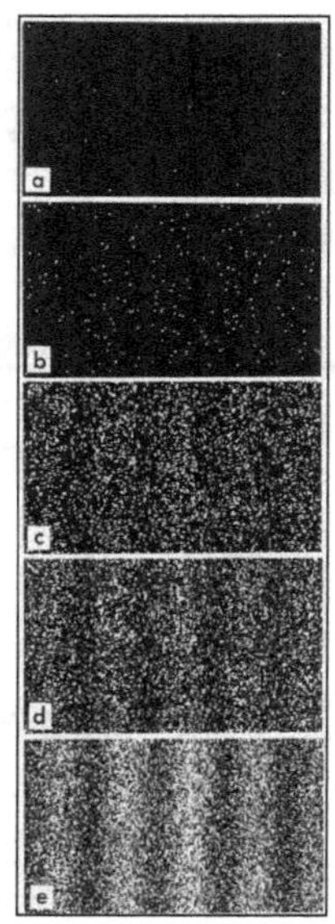

A double slit experiment showing the accumulation of electrons on a screen as time passes.

Two-slit diffraction is a famous example of the strange behaviors that waves regularly display, that are not intuitively associated with particles. The overlapping waves from the two slits cancel each other out in some locations, and reinforce each other in other locations, causing a complex pattern to emerge. Intuitively, one would not expect this pattern from firing a single particle at the slits, because the particle should pass through one slit or the other, not a complex overlap of both.

However, since the Schrödinger equation is a wave equation, a single particle fired through a double-slit does show this same pattern. The experiment must be repeated many times for the complex pattern to emerge. Although this is counterintuitive, the prediction is correct; in particular, electron diffraction and neutron diffraction are well understood and widely used in science and engineering.

Related to diffraction, particles also display superposition and interference.

The superposition property allows the particle to be in a quantum superposition of two or more quantum states at the same time. However, a "quantum state" in quantum mechanics means the probability that a system will be, for example at a position x, not that the system will actually be at position x. It does not imply that the particle itself may be in two classical states at once. Indeed, quantum mechanics is generally unable to assign values for properties prior to measurement at all.

Measurement and Uncertainty

In classical mechanics, a particle has, at every moment, an exact position and an exact momentum. These values change deterministically as the particle moves according to Newton's laws. Under the Copenhagen interpretation of quantum mechanics, particles do not have exactly determined properties, and when they are measured, the result is randomly drawn from a probability distribution. The Schrödinger equation predicts what the probability distributions are, but fundamentally cannot predict the exact result of each measurement.

The Heisenberg uncertainty principle is one statement of the inherent measurement uncertainty in quantum mechanics. It states that the more precisely a particle's position is known, the less precisely its momentum is known, and vice versa.

The Schrödinger equation describes the (deterministic) evolution of the wave function of a particle. However, even if the wave function is known exactly, the result of a specific measurement on the wave function is uncertain.

Interpretation of the Wave Function

The Schrödinger equation provides a way to calculate the wave function of a system and how it changes dynamically in time. However, the Schrödinger equation does not directly say what, exactly, the wave function is. Interpretations of quantum mechanics address questions such as what the relation is between the wave function, the underlying reality, and the results of experimental measurements.

An important aspect is the relationship between the Schrödinger equation and wave function collapse. In the oldest Copenhagen interpretation, particles follow the Schrödinger equation except during wave function collapse, during which they behave entirely differently. The advent of quantum decoherence theory allowed alternative approaches (such as the Everett many-worlds interpretation and consistent histories), wherein the Schrödinger equation is always satisfied, and wave function collapse should be explained as a consequence of the Schrödinger equation.

In 1952, Erwin Schrödinger gave a lecture during which he commented:

> Nearly every result a quantum theorist pronounces is about the probability of this or that or that happening—with usually a great many alternatives. The idea

> that they be not alternatives but all really happen simultaneously seems lunatic to him, just impossible.

David Deutsch regarded this as the earliest known reference to a many-worlds interpretation of quantum mechanics, an interpretation generally credited to Hugh Everett III, while Jeffrey A. Barrett took the more modest position that it indicates a "similarity in general views" between Schrödinger and Everett.

Wave Equation for Particles

The Schrödinger equation is a variation on the diffusion equation where the diffusion constant is imaginary. A spike of heat will decay in amplitude and spread out; however, because the imaginary i is the generator of rotations in the complex plane, a spike in the amplitude of a matter wave will also rotate in the complex plane over time. The solutions are therefore functions which describe wave-like motions. Wave equations in physics can normally be derived from other physical laws – the wave equation for mechanical vibrations on strings and in matter can be derived from Newton's laws, where the wave function represents the displacement of matter, and electromagnetic waves from Maxwell's equations, where the wave functions are electric and magnetic fields. The basis for Schrödinger's equation, on the other hand, is the energy of the system and a separate postulate of quantum mechanics: the wave function is a description of the system. The Schrödinger equation is therefore a new concept in itself; as Feynman put it:

> "Where did we get that equation from? Nowhere. It is not possible to derive it from anything you know. It came out of the mind of Schrödinger."
>
> —Richard Feynman

The foundation of the equation is structured to be a linear differential equation based on classical energy conservation, and consistent with the De Broglie relations. The solution is the wave function ψ, which contains all the information that can be known about the system. In the Copenhagen interpretation, the modulus of ψ is related to the probability the particles are in some spatial configuration at some instant of time. Solving the equation for ψ can be used to predict how the particles will behave under the influence of the specified potential and with each other.

The Schrödinger equation was developed principally from the De Broglie hypothesis, a wave equation that would describe particles, and can be constructed.

Consistency with Energy Conservation

The total energy E of a particle is the sum of kinetic energy T and potential energy V, this sum is also the frequent expression for the Hamiltonian H in classical mechanics:

$$E = T + V = H$$

Explicitly, for a particle in one dimension with position x, mass m and momentum p, and potential energy *V* which generally varies with position and time t:

$$E = \frac{p^2}{2m} + V(x,t) = H$$

For three dimensions, the position vector **r** and momentum vector **p** must be used:

$$E = \frac{\mathbf{p} \cdot \mathbf{p}}{2m} + V(\mathbf{r},t) = H$$

This formalism can be extended to any fixed number of particles: the total energy of the system is then the total kinetic energies of the particles, plus the total potential energy, again the Hamiltonian. However, there can be interactions between the particles (an N-body problem), so the potential energy V can change as the spatial configuration of particles changes, and possibly with time. The potential energy, in general, is not the sum of the separate potential energies for each particle, it is a function of all the spatial positions of the particles. Explicitly:

$$E = \sum_{n=1}^{N} \frac{\mathbf{p}_n \cdot \mathbf{p}_n}{2m_n} + V(\mathbf{r}_1, \mathbf{r}_2 \cdots \mathbf{r}_N, t) = H$$

Linearity

The simplest wave function is a plane wave of the form:

$$\Psi(\mathbf{r},t) = Ae^{i(\mathbf{k}\cdot\mathbf{r}-\omega t)}$$

where the A is the amplitude, **k** the wave vector, and ω the angular frequency, of the plane wave. In general, physical situations are not purely described by plane waves, so for generality the superposition principle is required; any wave can be made by superposition of sinusoidal plane waves. So if the equation is linear, a linear combination of plane waves is also an allowed solution. Hence a necessary and separate requirement is that the Schrödinger equation is a linear differential equation. For discrete **k** the sum is a superposition of plane waves:

$$\Psi(\mathbf{r},t) = \sum_{n=1}^{\infty} A_n e^{i(\mathbf{k}_n\cdot\mathbf{r}-\omega_n t)}$$

for some real amplitude coefficients A_n, and for continuous **k** the sum becomes an integral, the Fourier transform of a momentum space wave function:

$$\Psi(\mathbf{r},t) = \frac{1}{(\sqrt{2\pi})^3} \int \Phi(\mathbf{k}) e^{i(\mathbf{k}\cdot\mathbf{r}-\omega t)} d^3\mathbf{k}$$

where $d^3\mathbf{k} = dk_x dk_x dk_z$ is the differential volume element in **k**-space, and the integrals are taken over all **k**-space. The momentum wave function $\Phi(\mathbf{k})$ arises in the integrand since the position and momentum space wave functions are Fourier transforms of each other.

Consistency with the De Broglie Relations

Einstein's light quanta hypothesis states that the energy E of a quantum of light or photon is proportional to its frequency v (or angular frequency, $\omega = 2\pi\nu$):

$$E = h\nu = \hbar\omega$$

Likewise De Broglie's hypothesis states that any particle can be associated with a wave, and that the momentum p of the particle is inversely proportional to the wavelength λ of such a wave (or proportional to the wavenumber, $k = \frac{2\pi}{\lambda}$ in one dimension, by:

$$p = \frac{h}{\lambda} = \hbar k,$$

while in three dimensions, wavelength λ is related to the magnitude of the wavevector k:

$$\mathbf{p} = \hbar\mathbf{k}, \quad |\mathbf{k}| = \frac{2\pi}{\lambda}.$$

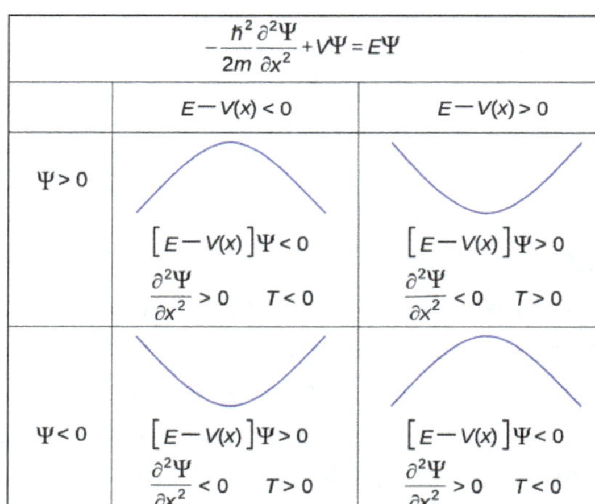

Diagrammatic summary of the quantities related to the wave function, as used in De broglie's hypothesis and development of the Schrödinger equation.

The Planck–Einstein and de Broglie relations illuminate the deep connections between energy with time, and space with momentum, and express wave–particle duality. In practice, natural units comprising ħ = 1 are used, as the De Broglie equations reduce to identities: allowing momentum, wave number, energy and frequency to be used interchangeably, to prevent duplication of quantities, and reduce the number of dimensions of related quantities. For familiarity SI units are still used.

Schrödinger's insight, late in 1925, was to express the phase of a plane wave as a complex phase factor using these relations:

$$\Psi = Ae^{i(\mathbf{k}\cdot\mathbf{r} - \omega t)} = Ae^{i(\mathbf{p}\cdot\mathbf{r} - Et)/\hbar}$$

and to realize that the first order partial derivatives with respect to space were:

$$\nabla\Psi = \frac{i}{\hbar}\mathbf{p}Ae^{i(\mathbf{p}\cdot\mathbf{r}-Et)/\hbar} = \frac{i}{\hbar}\mathbf{p}\Psi.$$

Taking partial derivatives with respect to time gives:

$$\frac{\partial\Psi}{\partial t} = -\frac{iE}{\hbar}Ae^{i(\mathbf{p}\cdot\mathbf{r}-Et)/\hbar} = -\frac{iE}{\hbar}\Psi.$$

Another postulate of quantum mechanics is that all observables are represented by linear Hermitian operators which act on the wave function, and the eigenvalues of the operator are the values the observable takes. The previous derivatives are consistent with the energy operator (or Hamiltonian operator), corresponding to the time derivative,

$$\hat{E}\Psi = i\hbar\frac{\partial}{\partial t}\Psi = E\Psi$$

where E are the energy eigenvalues, and the momentum operator, corresponding to the spatial derivatives (the gradient ∇).

$$\hat{\mathbf{p}}\Psi = -i\hbar\nabla\Psi = \mathbf{p}\Psi$$

where **p** is a vector of the momentum eigenvalues. In the above, the "hats" (^) indicate these observables are operators, not simply ordinary numbers or vectors. The energy and momentum operators are differential operators, while the potential energy operator V is just a multiplicative factor.

Substituting the energy and momentum operators into the classical energy conservation equation obtains the operator:

$$E = \frac{\mathbf{p}\cdot\mathbf{p}}{2m} + V \quad \rightarrow \quad \hat{E} = \frac{\hat{\mathbf{p}}\cdot\hat{\mathbf{p}}}{2m} + V$$

so in terms of derivatives with respect to time and space, acting this operator on the wave function Ψ immediately led Schrödinger to his equation:

$$i\hbar\frac{\partial\Psi}{\partial t} = -\frac{\hbar^2}{2m}\nabla^2\Psi + V\Psi$$

Wave–particle duality can be assessed from these equations as follows. The kinetic energy T is related to the square of momentum **p**. As the particle's momentum increases, the kinetic energy increases more rapidly, but since the wave number $|\mathbf{k}|$ increases the wavelength λ decreases. In terms of ordinary scalar and vector quantities (not operators):

$$\mathbf{p}\cdot\mathbf{p} \propto \mathbf{k}\cdot\mathbf{k} \propto T \propto \frac{1}{\lambda^2}$$

The kinetic energy is also proportional to the second spatial derivatives, so it is also proportional to the magnitude of the curvature of the wave, in terms of operators:

$$\hat{T}\Psi = \frac{-\hbar^2}{2m}\nabla\cdot\nabla\Psi \propto \nabla^2\Psi.$$

As the curvature increases, the amplitude of the wave alternates between positive and negative more rapidly, and also shortens the wavelength. So the inverse relation between momentum and wavelength is consistent with the energy the particle has, and so the energy of the particle has a connection to a wave, all in the same mathematical formulation.

Wave and Particle Motion

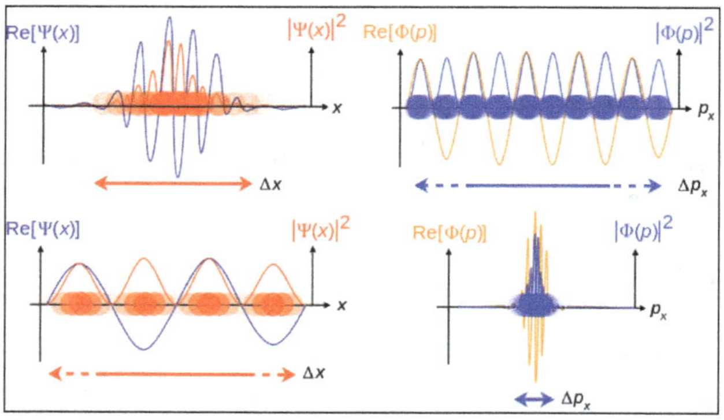

Increasing levels of wavepacket localization, meaning the particle has a more localized position.

Schrödinger required that a wave packet solution near position **r** with wave vector near **k** will move along the trajectory determined by classical mechanics for times short enough for the spread in **k** (and hence in velocity) not to substantially increase the spread in **r**. Since, for a given spread in **k**, the spread in velocity is proportional to Planck's constant $\hbar$, it is sometimes said that in the limit as $\hbar$ approaches zero, the equations of classical mechanics are restored from quantum mechanics. Great care is required in how that limit is taken, and in what cases.

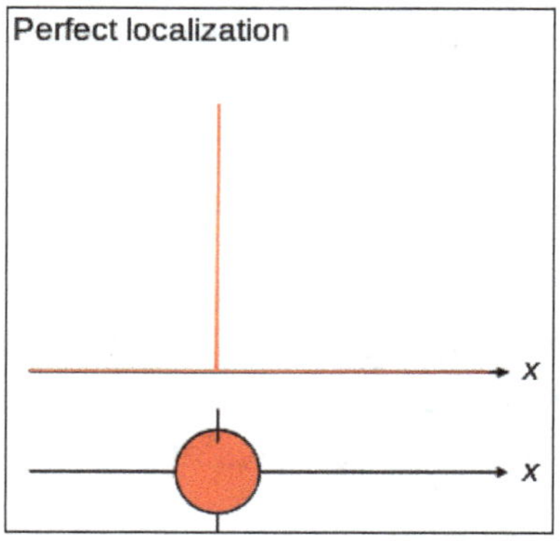

In the limit $\hbar \to 0$, the particle's position and momentum become known exactly. This is equivalent to the classical particle.

The limiting short-wavelength is equivalent to $\hbar$ tending to zero because this is limiting case of increasing the wave packet localization to the definite position of the particle. Using the Heisenberg uncertainty principle for position and momentum, the products of uncertainty in position and momentum become zero as $\hbar \to 0$:

$$\sigma(x)\sigma(p_x) \geq \frac{\hbar}{2} \quad \to \quad \sigma(x)\sigma(p_x) \geq 0$$

where σ denotes the (root mean square) measurement uncertainty in x and p_x (and similarly for the y and z directions) which implies the position and momentum can only be known to arbitrary precision in this limit.

One simple way to compare classical to quantum mechanics is to consider the time-evolution of the expected position and expected momentum, which can then be compared to the time-evolution of the ordinary position and momentum in classical mechanics. The quantum expectation values satisfy the Ehrenfest theorem. For a one-dimensional quantum particle moving in a potential V, the Ehrenfest theorem says,

$$m\frac{d}{dt}\langle x\rangle = \langle p\rangle; \quad \frac{d}{dt}\langle p\rangle = -\langle V'(X)\rangle.$$

Although the first of these equations is consistent with the classical behavior, the second is not: If the pair $(\langle X\rangle, \langle P\rangle)$ were to satisfy Newton's second law, the right-hand side of the second equation would have to be,

$$-V'(\langle X\rangle),$$

which is typically not the same as $-\langle V'(X)\rangle$ In the case of the quantum harmonic oscillator, however, V′ is linear and this distinction disappears, so that in this very special case, the expected position and expected momentum do exactly follow the classical trajectories.

For general systems, the best we can hope for is that the expected position and momentum will approximately follow the classical trajectories. If the wave function is highly concentrated around a point x_0 , then $V'(\langle X\rangle)$ and $\langle V'(X)\rangle$ will be almost the same, since both will be approximately equal to $V'(x_0)$. In that case, the expected position and expected momentum will remain very close to the classical trajectories, at least for as long as the wave function remains highly localized in position. When Planck's constant is small, it is possible to have a state that is well localized in both position and momentum. The small uncertainty in momentum ensures that the particle remains well localized in position for a long time, so that expected position and momentum continue to closely track the classical trajectories.

The Schrödinger equation in its general form,

$$i\hbar\frac{\partial}{\partial t}\Psi(\mathbf{r},t) = \hat{H}\Psi(\mathbf{r},t)$$

is closely related to the Hamilton–Jacobi equation (HJE),

$$-\frac{\partial}{\partial t} S(q_i,t) = H\left(q_i, \frac{\partial S}{\partial q_i}, t\right)$$

where S is the classical action and H is the Hamiltonian function (not operator). Here the generalized coordinates q_i for $i = 1,2,3$ (used in the context of the HJE) can be set to the position in Cartesian coordinates as $\mathbf{r} = (q_1, q_2, q_3) = (x, y, z)$.

Substituting,

$$\Psi = \sqrt{\rho(\mathbf{r},t)} e^{iS(\mathbf{r},t)/\hbar}$$

where ρ is the probability density, into the Schrödinger equation and then taking the limit $\hbar \to 0$ in the resulting equation yield the Hamilton–Jacobi equation. The implications are as follows:

- The motion of a particle, described by a (short-wavelength) wave packet solution to the Schrödinger equation, is also described by the Hamilton–Jacobi equation of motion.
- The Schrödinger equation includes the wave function, so its wave packet solution implies the position of a (quantum) particle is fuzzily spread out in wave fronts. On the contrary, the Hamilton–Jacobi equation applies to a (classical) particle of definite position and momentum, instead the position and momentum at all times (the trajectory) are deterministic and can be simultaneously known.

Nonrelativistic Quantum Mechanics

The quantum mechanics of particles without accounting for the effects of special relativity, for example particles propagating at speeds much less than light, is known as nonrelativistic quantum mechanics. Following are several forms of Schrödinger's equation in this context for different situations: time independence and dependence, one and three spatial dimensions, and one and N particles.

In actuality, the particles constituting the system do not have the numerical labels used in theory. The language of mathematics forces us to label the positions of particles one way or another, otherwise there would be confusion between symbols representing which variables are for which particle.

Time-independent

If the Hamiltonian is not an explicit function of time, the equation is separable into a product of spatial and temporal parts. In general, the wave function takes the form:

$$\Psi(\text{space coords}, t) = \psi(\text{space coords})\tau(t).$$

where ψ(space coords) is a function of all the spatial coordinates of the particles constituting the system only, and τ(t) is a function of time only.

Substituting for ψ into the Schrödinger equation for the relevant number of particles in the relevant number of dimensions, solving by separation of variables implies the general solution of the time-dependent equation has the form:

$$\Psi(\text{space coords}, t) = \psi(\text{space coords})e^{-iEt/\hbar}.$$

Since the time dependent phase factor is always the same, only the spatial part needs to be solved for in time independent problems. Additionally, the energy operator $\hat{E} = i\hbar\frac{\partial}{\partial t}$ can always be replaced by the energy eigenvalue E, thus the time independent Schrödinger equation is an eigenvalue equation for the Hamiltonian operator:

$$\hat{H}\psi = E\psi$$

This is true for any number of particles in any number of dimensions (in a time independent potential). This case describes the standing wave solutions of the time-dependent equation, which are the states with definite energy (instead of a probability distribution of different energies). In physics, these standing waves are called "stationary states" or "energy eigenstates"; in chemistry they are called "atomic orbitals" or "molecular orbitals". Superpositions of energy eigenstates change their properties according to the relative phases between the energy levels.

The energy eigenvalues from this equation form a discrete spectrum of values, so mathematically energy must be quantized. More specifically, the energy eigenstates form a basis – any wave function may be written as a sum over the discrete energy states or an integral over continuous energy states, or more generally as an integral over a measure. This is the spectral theorem in mathematics, and in a finite state space it is just a statement of the completeness of the eigenvectors of a Hermitian matrix.

One-dimensional Examples

For a particle in one dimension, the Hamiltonian is:

$$\hat{H} = \frac{\hat{p}^2}{2m} + V(x), \quad \hat{p} = -i\hbar\frac{d}{dx}$$

and substituting this into the general Schrödinger equation gives:

$$\left[-\frac{\hbar^2}{2m}\frac{d^2}{dx^2} + V(x)\right]\psi(x) = E\psi(x)$$

This is the only case the Schrödinger equation is an ordinary differential equation, rather than a partial differential equation. The general solutions are always of the form:

$$\Psi(x,t) = \psi(x)e^{-iEt/\hbar}.$$

For *N* particles in one dimension, the Hamiltonian is:

$$\hat{H} = \sum_{n=1}^{N} \frac{\hat{p}_n^2}{2m_n} + V(x_1, x_2, \cdots x_N), \quad \hat{p}_n = -i\hbar \frac{\partial}{\partial x_n}$$

where the position of particle n is x_n. The corresponding Schrödinger equation is:

$$-\frac{\hbar^2}{2}\sum_{n=1}^{N}\frac{1}{m_n}\frac{\partial^2}{\partial x_n^2}\psi(x_1,x_2,\cdots x_N) + V(x_1,x_2,\cdots x_N)\psi(x_1,x_2,\cdots x_N) = E\psi(x_1,x_2,\cdots x_N).$$

so the general solutions have the form:

$$\Psi(x_1, x_2, \cdots x_N, t) = e^{-iEt/\hbar}\psi(x_1, x_2 \cdots x_N)$$

For non-interacting distinguishable particles, the potential of the system only influences each particle separately, so the total potential energy is the sum of potential energies for each particle:

$$V(x_1, x_2, \cdots x_N) = \sum_{n=1}^{N} V(x_n).$$

and the wave function can be written as a product of the wave functions for each particle:

$$\Psi(x_1, x_2, \cdots x_N, t) = e^{-iEt/\hbar}\prod_{n=1}^{N}\psi(x_n),$$

For non-interacting identical particles, the potential is still a sum, but wave function is a bit more complicated – it is a sum over the permutations of products of the separate wave functions to account for particle exchange. In general for interacting particles, the above decompositions are not possible.

Free Particle

For no potential, V = 0, so the particle is free and the equation reads:

$$-E\psi = \frac{\hbar^2}{2m}\frac{d^2\psi}{dx^2}$$

which has oscillatory solutions for E > 0 (the C_n are arbitrary constants):

$$\psi_E(x) = C_1 e^{i\sqrt{2mE/\hbar^2}\,x} + C_2 e^{-i\sqrt{2mE/\hbar^2}\,x}$$

and exponential solutions for E < 0:

$$\psi_{-|E|}(x) = C_1 e^{\sqrt{2m|E|/\hbar^2}\, x} + C_2 e^{-\sqrt{2m|E|/\hbar^2}\, x}.$$

The exponentially growing solutions have an infinite norm, and are not physical. They are not allowed in a finite volume with periodic or fixed boundary conditions.

Constant Potential

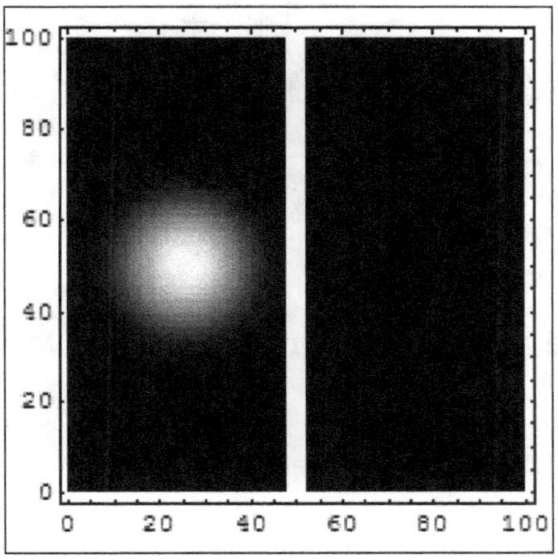

For a constant potential, $V = V_0$, the solution is oscillatory for $E > V_0$ and exponential for $E < V_0$, corresponding to energies that are allowed or disallowed in classical mechanics. Oscillatory solutions have a classically allowed energy and correspond to actual classical motions, while the exponential solutions have a disallowed energy and describe a small amount of quantum bleeding into the classically disallowed region, due to quantum tunneling. If the potential V_0 grows to infinity, the motion is classically confined to a finite region. Viewed far enough away, every solution is reduced to an exponential; the condition that the exponential is decreasing restricts the energy levels to a discrete set, called the allowed energies.

Harmonic Oscillator

The Schrödinger equation for this situation is:

$$E\psi = -\frac{\hbar^2}{2m}\frac{d^2}{dx^2}\psi + \frac{1}{2}m\omega^2 x^2 \psi,$$

where x is the displacement and ω the angular frequency. This is an example of a quantum-mechanical system whose wave function can be solved for exactly. Furthermore, it can be used to describe approximately a wide variety of other systems, including vibrating atoms, molecules, and atoms or ions in lattices, and approximating other potentials near equilibrium points. It is also the basis of perturbation methods in quantum mechanics.

The solutions in position space are:

$$\psi_n(x)=\sqrt{\frac{1}{2^n\,n!}}\cdot\left(\frac{m\omega}{\pi\hbar}\right)^{1/4}\cdot e^{-\frac{m\omega x^2}{2\hbar}}\cdot\mathcal{H}_n\left(\sqrt{\frac{m\omega}{\hbar}}x\right),$$

where $n\in\{0,1,2,\ldots\}n\in\{0,1,2,\ldots\}$, and the functions $\mathcal{H}_n$ are the Hermite polynomials of order n. The solution set may be generated by:

$$\psi_n(x)=\frac{1}{\sqrt{n!}}\left(\sqrt{\frac{m\omega}{2\hbar}}\right)^n\left(x-\frac{\hbar}{m\omega}\frac{d}{dx}\right)^n\left(\frac{m\omega}{\pi\hbar}\right)^{\frac{1}{4}}e^{\frac{-m\omega x^2}{2\hbar}}$$

The eigenvalues are:

$$E_n=\left(n+\frac{1}{2}\right)\hbar\omega.$$

The case n = 0 is called the zero-point energy and the wave function is a Gaussian.

Three-dimensional Examples

The extension from one dimension to three dimensions is straightforward, all position and momentum operators are replaced by their three-dimensional expressions and the partial derivative with respect to space is replaced by the gradient operator.

The Hamiltonian for one particle in three dimensions is:

$$\hat{H}=\frac{\hat{\mathbf{p}}\cdot\hat{\mathbf{p}}}{2m}+V(\mathbf{r}),\quad \hat{\mathbf{p}}=-i\hbar\nabla$$

generating the equation:

$$\left[-\frac{\hbar^2}{2m}\nabla^2+V(\mathbf{r})\right]\psi(\mathbf{r})=E\psi(\mathbf{r})$$

with stationary state solutions of the form,

$$\Psi(\mathbf{r},t)=\psi(\mathbf{r})e^{-iEt/\hbar},$$

where the position of the particle is **r**.

For N particles in three dimensions, the Hamiltonian is:

$$\hat{H}=\sum_{n=1}^{N}\frac{\hat{\mathbf{p}}_n\cdot\hat{\mathbf{p}}_n}{2m_n}+V(\mathbf{r}_1,\mathbf{r}_2,\cdots\mathbf{r}_N),\quad \hat{\mathbf{p}}_n=-i\hbar\nabla_n$$

where the position of particle n is $\mathbf{r}_n$ and the gradient operators are partial derivatives with respect to the particle's position coordinates. In Cartesian coordinates, for particle n, the position vector is $\mathbf{r}_n = (x_n, y_n, z_n)$ while the gradient and Laplacian operator are respectively:

$$\nabla_n = \mathbf{e}_x \frac{\partial}{\partial x_n} + \mathbf{e}_y \frac{\partial}{\partial y_n} + \mathbf{e}_z \frac{\partial}{\partial z_n}, \quad \nabla_n^2 = \nabla_n \cdot \nabla_n = \frac{\partial^2}{\partial {x_n}^2} + \frac{\partial^2}{\partial {y_n}^2} + \frac{\partial^2}{\partial {z_n}^2}$$

The Schrödinger equation is:

$$-\frac{\hbar^2}{2} \sum_{n=1}^{N} \frac{1}{m_n} \nabla_n^2 \Psi(\mathbf{r}_1, \mathbf{r}_2, \cdots \mathbf{r}_N) + V(\mathbf{r}_1, \mathbf{r}_2, \cdots \mathbf{r}_N) \Psi(\mathbf{r}_1, \mathbf{r}_2, \cdots \mathbf{r}_N) = E \Psi(\mathbf{r}_1, \mathbf{r}_2, \cdots \mathbf{r}_N)$$

with stationary state solutions:

$$\Psi(\mathbf{r}_1, \mathbf{r}_2 \cdots \mathbf{r}_N, t) = e^{-iEt/\hbar} \psi(\mathbf{r}_1, \mathbf{r}_2 \cdots \mathbf{r}_N)$$

Again, for non-interacting distinguishable particles the potential is the sum of particle potentials,

$$V(\mathbf{r}_1, \mathbf{r}_2, \cdots \mathbf{r}_N) = \sum_{n=1}^{N} V(\mathbf{r}_n)$$

and the wave function is a product of the particle wave functions,

$$\Psi(\mathbf{r}_1, \mathbf{r}_2 \cdots \mathbf{r}_N, t) = e^{-iEt/\hbar} \prod_{n=1}^{N} \psi(\mathbf{r}_n).$$

For non-interacting identical particles, the potential is a sum but the wave function is a sum over permutations of products. The previous two equations do not apply to interacting particles.

Following are examples where exact solutions are known.

Hydrogen Atom

The Schrödinger equation the hydrogen atom (or a hydrogen-like atom) is:

$$E\psi = -\frac{\hbar^2}{2\mu} \nabla^2 \psi - \frac{q^2}{4\pi\varepsilon_0 r} \psi$$

where q is the electron charge, **r** is the position of the electron relative to the nucleus, **r** = | **r** | is the magnitude of the relative position, the potential term is due to the Coulomb interaction, wherein ε_0 is the permittivity of free space:

$$\mu = \frac{m_q m_p}{m_q + m_p}$$

is the 2-body reduced mass of the hydrogen nucleus (just a proton) of mass m_p and the electron of mass m_q. The negative sign arises in the potential term since the proton and electron are oppositely charged. The reduced mass in place of the electron mass is used since the electron and proton together orbit each other about a common centre of mass, and constitute a two-body problem to solve. The motion of the electron is of principle interest here, so the equivalent one-body problem is the motion of the electron using the reduced mass.

The Schrödinger for a hydrogen atom can be solved by separation of variables. In this case, spherical polar coordinates are the most convenient. Thus:

$$\psi(r,\theta,\phi) = R(r)Y_\ell^m(\theta,\phi) = R(r)\Theta(\theta)\Phi(\phi),$$

where R are radial functions and $Y_l^m(\theta,\phi)$ are spherical harmonics of degree ℓ and order m. This is the only atom for which the Schrödinger equation has been solved for exactly. Multi-electron atoms require approximate methods. The family of solutions are:

$$\psi_{n\ell m}(r,\theta,\phi) = \sqrt{\left(\frac{2}{na_0}\right)^3 \frac{(n-\ell-1)!}{2n[(n+\ell)!]}} e^{-r/na_0} \left(\frac{2r}{na_0}\right)^\ell L_{n-\ell-1}^{2\ell+1}\left(\frac{2r}{na_0}\right) \cdot Y_\ell^m(\theta,\phi)$$

where:

- $a_0 = \dfrac{4\pi\varepsilon_0\hbar^2}{m_q q^2}$ is the Bohr radius.
- $L_{n-\ell-1}^{2\ell+1}(\cdots)$ are the generalized Laguerre polynomials of degree $n-\ell-1$.
- n,ℓ,m are the principal, azimuthal, and magnetic quantum numbers respectively, which take the values:

  $n = 1,2,3,\ldots$

  $\ell = 0,1,2,\ldots,n-1$

  $m = -\ell,\ldots,\ell$

The generalized Laguerre polynomials are defined differently by different authors.

Two-electron Atoms or Ions

The equation for any two-electron system, such as the neutral helium atom (He, $Z=2$), the negative hydrogen ion (H^-, $Z=1$), or the positive lithium ion (Li^+, $Z=3$) is:

$$E\psi = -\hbar^2\left[\frac{1}{2\mu}\left(\nabla_1^2+\nabla_2^2\right)+\frac{1}{M}\nabla_1\cdot\nabla_2\right]\psi + \frac{e^2}{4\pi\varepsilon_0}\left[\frac{1}{r_{12}} - Z\left(\frac{1}{r_1}+\frac{1}{r_2}\right)\right]\psi$$

where r_1 is the relative position of one electron ($r_1 = |\mathbf{r}_1|$ is its relative magnitude), $\mathbf{r}_2$ is the relative position of the other electron ($r_2 = |\mathbf{r}_2|$ is the magnitude), $r_{12} = |\mathbf{r}_{12}|$ is the magnitude of the separation between them given by:

$$|\mathbf{r}_{12}| = |\mathbf{r}_2 - \mathbf{r}_1|$$

μ is again the two-body reduced mass of an electron with respect to the nucleus of mass M, so this time:

$$\mu = \frac{m_e M}{m_e + M}$$

and Z is the atomic number for the element (not a quantum number).

The cross-term of two Laplacians:

$$\frac{1}{M}\nabla_1 \cdot \nabla_2$$

is known as the mass polarization term, which arises due to the motion of atomic nuclei. The wave function is a function of the two electron's positions:

$$\psi = \psi(\mathbf{r}_1, \mathbf{r}_2)$$

There is no closed form solution for this equation.

Time-dependent

This is the equation of motion for the quantum state. In the most general form, it is written:

$$i\hbar \frac{\partial}{\partial t}\Psi = \hat{H}\Psi.$$

and the solution, the wave function, is a function of all the particle coordinates of the system and time. Following are specific cases.

For one particle in one dimension, the Hamiltonian:

$$\hat{H} = \frac{\hat{p}^2}{2m} + V(x,t), \quad \hat{p} = -i\hbar\frac{\partial}{\partial x}$$

generates the equation:

$$i\hbar\frac{\partial}{\partial t}\Psi(x,t) = -\frac{\hbar^2}{2m}\frac{\partial^2}{\partial x^2}\Psi(x,t) + V(x,t)\Psi(x,t)$$

For N particles in one dimension, the Hamiltonian is:

$$\hat{H}=\sum_{n=1}^{N}\frac{\hat{p}_n^2}{2m_n}+V(x_1,x_2,\cdots x_N,t),\quad \hat{p}_n=-i\hbar\frac{\partial}{\partial x_n}$$

where the position of particle n is x_n, generating the equation:

$$i\hbar\frac{\partial}{\partial t}\Psi(x_1,x_2\cdots x_N,t)=-\frac{\hbar^2}{2}\sum_{n=1}^{N}\frac{1}{m_n}\frac{\partial^2}{\partial x_n^2}\Psi(x_1,x_2\cdots x_N,t)+V(x_1,x_2\cdots x_N,t)\Psi(x_1,x_2\cdots x_N,t).$$

For one particle in three dimensions, the Hamiltonian is:

$$\hat{H}=\frac{\hat{\mathbf{p}}\cdot\hat{\mathbf{p}}}{2m}+V(\mathbf{r},t),\quad \hat{\mathbf{p}}=-i\hbar\nabla$$

generating the equation:

$$i\hbar\frac{\partial}{\partial t}\Psi(\mathbf{r},t)=-\frac{\hbar^2}{2m}\nabla^2\Psi(\mathbf{r},t)+V(\mathbf{r},t)\Psi(\mathbf{r},t)$$

For N particles in three dimensions, the Hamiltonian is:

$$\hat{H}=\sum_{n=1}^{N}\frac{\hat{\mathbf{p}}_n\cdot\hat{\mathbf{p}}_n}{2m_n}+V(\mathbf{r}_1,\mathbf{r}_2,\cdots\mathbf{r}_N,t),\quad \hat{\mathbf{p}}_n=-i\hbar\nabla_n$$

where the position of particle n is r_n, generating the equation:

$$i\hbar\frac{\partial}{\partial t}\Psi(\mathbf{r}_1,\mathbf{r}_2,\cdots\mathbf{r}_N,t)=-\frac{\hbar^2}{2}\sum_{n=1}^{N}\frac{1}{m_n}\nabla_n^2\Psi(\mathbf{r}_1,\mathbf{r}_2,\cdots\mathbf{r}_N,t)+V(\mathbf{r}_1,\mathbf{r}_2,\cdots\mathbf{r}_N,t)\Psi(\mathbf{r}_1,\mathbf{r}_2,\cdots\mathbf{r}_N,t)$$

This last equation is in a very high dimension, so the solutions are not easy to visualize.

Properties

The Schrödinger equation has the following properties: some are useful, but there are shortcomings. Ultimately, these properties arise from the Hamiltonian used, and the solutions to the equation.

Linearity

In the development above, the Schrödinger equation was made to be linear for generality, though this has other implications. If two wave functions ψ_1 and ψ_2 are solutions, then so is any linear combination of the two:

$$\psi=a\psi_1+b\psi_2$$

where a and b are any complex numbers (the sum can be extended for any number of wave functions). This property allows superpositions of quantum states to be solutions of the Schrödinger equation. Even more generally, it holds that a general solution to the Schrödinger equation can be found by taking a weighted sum over all single state solutions achievable. For example, consider a wave function $\Psi(x, t)$ such that the wave function is a product of two functions: one time independent, and one time dependent. If states of definite energy found using the time independent Schrödinger equation are given by $\psi_E(x)$ with amplitude A_n and time dependent phase factor is given by:

$$e^{-iE_n t/\hbar},$$

then a valid general solution is:

$$\Psi(x,t) = \sum_n A_n \psi_{E_n}(x) e^{-iE_n t/\hbar}.$$

Additionally, the ability to scale solutions allows one to solve for a wave function without normalizing it first. If one has a set of normalized solutions ψ_n, then,

$$\Psi = \sum_n A_n \psi_n$$

can be normalized by ensuring that,

$$\sum_n |A_n|^2 = 1.$$

This is much more convenient than having to verify that,

$$\int_{-\infty}^{\infty} |\Psi(x)|^2 \, dx = \int_{-\infty}^{\infty} \Psi(x)\Psi^*(x) \, dx = 1.$$

Momentum Space Schrödinger Equation

The Schrödinger equation $i\hbar\partial_t |\psi\rangle = \hat{H} |\psi\rangle$ is often presented in the position basis form $i\hbar\partial_t \psi(x) = -\frac{\hbar^2}{2m}\nabla^2\psi(x) + V(x)\psi(x)$ (with $\langle x|\psi\rangle \equiv \psi(x)$. But as a vector operator equation it has a valid representation in any arbitrary complete basis of kets in Hilbert space. For example, in the momentum space basis the equation reads,

$$i\hbar\partial_t f(p) = \frac{p^2}{2m} f(p) + (\tilde{V} * f)(p)$$

where $|p\rangle$ is the plane wave state of definite momentum p, $\langle p|\psi\rangle \equiv f(p) = \int \psi(x) e^{-ipx} dx$, $\tilde{V}$ is the Fourier transform of V, and $*$ denotes Fourier convolution.

In the 1D example with absence of a potential, $\tilde{V}=0$ (or similarly $\tilde{V}(k)\propto\delta(k)$ in the case of a background potential constant throughout space), each stationary state of energy $\hbar\omega=q^2/2m$ is of the form $f_q(p)=(c_+\delta(p-q)+c_-\delta(p+q))e^{-i\omega t}$.

for arbitrary complex coefficients $c_\pm$. Such a wave function, as expected in free space, is a superposition of plane waves moving right and left with momenta $\pm q$; upon momentum measurement the state would collapse to one of definite momentum $\pm q$ with probability $\propto|c_\pm|^2$.

A version of the momentum space Schrödinger equation is often used in solid-state physics, as Bloch's theorem ensures the periodic crystal lattice potential couples f(p) with f(p + K) for only discrete reciprocal lattice vectors K. This makes it convenient to solve the momentum space Schrödinger equation at each point in the Brillouin zone independently of the other points in the Brillouin zone.

Real Energy Eigenstates

For the time-independent equation, an additional feature of linearity follows: if two wave functions ψ_1 and ψ_2 are solutions to the time-independent equation with the same energy E, then so is any linear combination:

$$\hat{H}(a\psi_1+b\psi_2)=a\hat{H}\psi_1+b\hat{H}\psi_2=E(a\psi_1+b\psi_2)$$

Two different solutions with the same energy are called degenerate.

In an arbitrary potential, if a wave function ψ solves the time-independent equation, so does its complex conjugate, denoted ψ*. By taking linear combinations, the real and imaginary parts of ψ are each solutions. If there is no degeneracy they can only differ by a factor.

In the time-dependent equation, complex conjugate waves move in opposite directions. If Ψ(x, t) is one solution, then so is Ψ*(x, –t). The symmetry of complex conjugation is called time-reversal symmetry.

Space and Time Derivatives

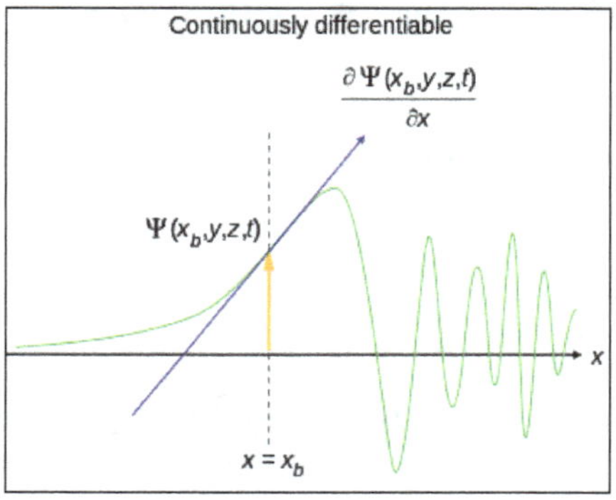

The Schrödinger equation is first order in time and second in space, which describes the time evolution of a quantum state (meaning it determines the future amplitude from the present).

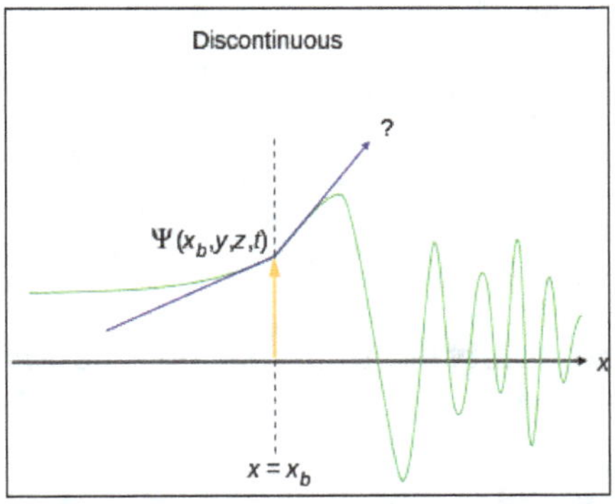

Continuity of the wave function and its first spatial derivative (in the x direction, y and z coordinates not shown), at some time t.

Explicitly for one particle in 3-dimensional Cartesian coordinates – the equation is:

$$i\hbar\frac{\partial\Psi}{\partial t}=-\frac{\hbar^2}{2m}\left(\frac{\partial^2\Psi}{\partial x^2}+\frac{\partial^2\Psi}{\partial y^2}+\frac{\partial^2\Psi}{\partial z^2}\right)+V(x,y,z,t)\Psi.$$

The first time partial derivative implies the initial value (at t = 0) of the wave function:

$$\Psi(x,y,z,0)$$

is an arbitrary constant. Likewise – the second order derivatives with respect to space implies the wave function and its first order spatial derivatives:

$$\Psi(x_b,y_b,z_b,t)$$

$$\frac{\partial}{\partial x}\Psi(x_b,y_b,z_b,t)\quad \frac{\partial}{\partial y}\Psi(x_b,y_b,z_b,t)\quad \frac{\partial}{\partial z}\Psi(x_b,y_b,z_b,t),$$

are all arbitrary constants at a given set of points, where x_b, y_b, z_b are a set of points describing boundary b (derivatives are evaluated at the boundaries). Typically there are one or two boundaries, such as the step potential and particle in a box respectively.

As the first order derivatives are arbitrary, the wave function can be a continuously differentiable function of space, since at any boundary the gradient of the wave function can be matched.

On the contrary, wave equations in physics are usually *second order in time*, notable are the family of classical wave equations and the quantum Klein–Gordon equation.

Local Conservation of Probability

The Schrödinger equation is consistent with probability conservation. Multiplying the

Schrödinger equation on the right by the complex conjugate wave function, and multiplying the wave function to the left of the complex conjugate of the Schrödinger equation, and subtracting, gives the continuity equation for probability:

$$\frac{\partial}{\partial t}\rho(\mathbf{r},t)+\nabla\cdot\mathbf{j}=0,$$

where:

$$\rho=|\Psi|^2=\Psi^*(\mathbf{r},t)\Psi(\mathbf{r},t)$$

is the probability density (probability per unit volume, * denotes complex conjugate), and,

$$\mathbf{j}=\frac{1}{2m}\left(\Psi^*\hat{\mathbf{p}}\Psi-\Psi\hat{\mathbf{p}}\Psi^*\right)$$

is the probability current (flow per unit area).

Hence predictions from the Schrödinger equation do not violate probability conservation.

Positive Energy

If the potential is bounded from below, meaning there is a minimum value of potential energy, the eigenfunctions of the Schrödinger equation have energy which is also bounded from below. This can be seen most easily by using the variational principle, as follows.

For any linear operator Â bounded from below, the eigenvector with the smallest eigenvalue is the vector ψ that minimizes the quantity:

$$\langle\psi|\hat{A}|\psi\rangle$$

over all ψ which are normalized. In this way, the smallest eigenvalue is expressed through the variational principle. For the Schrödinger Hamiltonian Ĥ bounded from below, the smallest eigenvalue is called the ground state energy. That energy is the minimum value of:

$$\langle\psi|\hat{H}|\psi\rangle=\int\psi^*(\mathbf{r})\left[-\frac{\hbar^2}{2m}\nabla^2\psi(\mathbf{r})+V(\mathbf{r})\psi(\mathbf{r})\right]d^3\mathbf{r}$$

$$=\int\left[\frac{\hbar^2}{2m}|\nabla\psi|^2+V(\mathbf{r})|\psi|^2\right]d^3\mathbf{r}=\langle\hat{H}\rangle$$

(using integration by parts). Due to the complex modulus of ψ^2 (which is positive definite), the right hand side is always greater than the lowest value of V(x). In particular, the ground state energy is positive when V(x) is everywhere positive.

For potentials which are bounded below and are not infinite over a region, there is a ground state which minimizes the integral above. This lowest energy wave function is real and positive definite – meaning the wave function can increase and decrease, but is positive for all positions. It physically cannot be negative: if it were, smoothing out the bends at the sign change (to minimize the wave function) rapidly reduces the gradient contribution to the integral and hence the kinetic energy, while the potential energy changes linearly and less quickly.

The kinetic and potential energy are both changing at different rates, so the total energy is not constant, which can't happen (conservation). The solutions are consistent with Schrödinger equation if this wave function is positive definite.

The lack of sign changes also shows that the ground state is nondegenerate, since if there were two ground states with common energy E, not proportional to each other, there would be a linear combination of the two that would also be a ground state resulting in a zero solution.

Analytic Continuation to Diffusion

The above properties (positive definiteness of energy) allow the analytic continuation of the Schrödinger equation to be identified as a stochastic process. This can be interpreted as the Huygens–Fresnel principle applied to De Broglie waves; the spreading wavefronts are diffusive probability amplitudes. For a free particle (not subject to a potential) in a random walk, substituting $\tau = it$ into the time-dependent Schrödinger equation gives:

$$\frac{\partial}{\partial \tau} X(\mathbf{r}, \tau) = \frac{\hbar}{2m} \nabla^2 X(\mathbf{r}, \tau), \quad X(\mathbf{r}, \tau) = \Psi(\mathbf{r}, \tau / i)$$

which has the same form as the diffusion equation, with diffusion coefficient $\frac{\hbar}{2m}$.

Regularity

On the space L^2 of square-integrable densities, the Schrödinger semigroup $e^{it\hat{H}}$ is a unitary evolution, and therefore surjective. The flows satisfy the Schrödinger equation $i\partial_t u = \hat{H}u$, where the derivative is taken in the distribution sense. However, since $\hat{H}$ for most physically reasonable Hamiltonians (e.g., the Laplace operator, possibly modified by a potential) is unbounded in L^2, this shows that the semigroup flows lack Sobolev regularity in general. Instead, solutions of the Schrödinger equation satisfy a Strichartz estimate.

Relativistic Quantum Mechanics

Relativistic quantum mechanics is obtained where quantum mechanics and special relativity simultaneously apply. In general, one wishes to build relativistic wave equations from the relativistic energy–momentum relation:

$$E^2 = (pc)^2 + (m_0c^2)^2,$$

instead of classical energy equations. The Klein–Gordon equation and the Dirac equation are two such equations. The Klein–Gordon equation:

$$\frac{1}{c^2}\frac{\partial^2}{\partial t^2}\psi - \nabla^2\psi + \frac{m^2c^2}{\hbar^2}\psi = 0,$$

was the first such equation to be obtained, even before the nonrelativistic one, and applies to massive spinless particles. The Dirac equation arose from taking the "square root" of the Klein–Gordon equation by factorizing the entire relativistic wave operator into a product of two operators – one of these is the operator for the entire Dirac equation. Entire Dirac equation:

$$\left(\beta mc^2 + c\left(\sum_{n=1}^{3}\alpha_n p_n\right)\right)\psi = i\hbar\frac{\partial\psi}{\partial t}$$

The general form of the Schrödinger equation remains true in relativity, but the Hamiltonian is less obvious. For example, the Dirac Hamiltonian for a particle of mass m and electric charge q in an electromagnetic field (described by the electromagnetic potentials φ and **A**) is:

$$\hat{H}_{\text{Dirac}} = \gamma^0\left[c\gamma\cdot\left(\hat{\mathbf{p}} - q\mathbf{A}\right) + mc^2 + \gamma^0 q\phi\right],$$

in which the $\gamma = (\gamma^1, \gamma^2, \gamma^3)$ and γ^0 are the Dirac gamma matrices related to the spin of the particle. The Dirac equation is true for all spin-½ particles, and the solutions to the equation are 4-component spinor fields with two components corresponding to the particle and the other two for the antiparticle.

For the Klein–Gordon equation, the general form of the Schrödinger equation is inconvenient to use, and in practice the Hamiltonian is not expressed in an analogous way to the Dirac Hamiltonian. The equations for relativistic quantum fields can be obtained in other ways, such as starting from a Lagrangian density and using the Euler–Lagrange equations for fields, or use the representation theory of the Lorentz group in which certain representations can be used to fix the equation for a free particle of given spin (and mass).

In general, the Hamiltonian to be substituted in the general Schrödinger equation is not just a function of the position and momentum operators (and possibly time), but also of spin matrices. Also, the solutions to a relativistic wave equation, for a massive particle of spin s, are complex-valued $2(2s + 1)$-component spinor fields.

EXPECTATION VALUES

To relate a quantum mechanical calculation to something you can observe in the laboratory, the "expectation value" of the measurable parameter is calculated. For the position x, the expectation value is defined as:

$$\langle x \rangle = \int_{-\infty}^{\infty} \Psi^*(x,t)\, x \Psi(x,t)\, dx$$

This integral can be interpreted as the average value of x that we would expect to obtain from a large number of measurements. Alternatively it could be viewed as the average value of position for a large number of particles which are described by the same wavefunction. For example, the expectation value of the radius of the electron in the ground state of the hydrogen atom is the average value you expect to obtain from making the measurement for a large number of hydrogen atoms.

While the expectation value of a function of position has the appearance of an average of the function, the expectation value of momentum involves the representation of momentum as a quantum mechanical operator.

$$\langle p \rangle = \int_{-\infty}^{\infty} \Psi^*(x,t) \frac{\hbar}{i} \frac{\partial}{\partial x} \Psi(x,t)\, dx$$

where:

$$P_{operator} = \frac{\hbar}{i} \frac{\partial}{\partial x}$$

is the operator for the x component of momentum.

Since the energy of a free particle is given by:

$$E = \frac{p^2}{2m} \quad then \ \langle E \rangle = \frac{\langle p^2 \rangle}{2m}$$

and the expectation value for energy becomes:

$$\langle E \rangle_{free\ particle} = \int_{-\infty}^{\infty} \Psi^* \frac{(-\hbar^2)}{2m} \frac{\partial^2}{\partial x^2} \Psi dx$$

for a particle in one dimension.

In general, the expectation value for any observable quantity is found by putting the quantum mechanical operator for that observable in the integral of the wavefunction over space:

$$\langle Q \rangle = \int_{-\infty}^{\infty} \Psi^* Q_{operator} \underset{\substack{\text{integral over} \\ \text{all space}}}{\Psi dV}$$

References

- Physics for Scientists and Engineers – with Modern Physics (6th Edition), P. A. Tipler, G. Mosca, Freeman, 2008, ISBN 0-7167-8964-7
- Wave-function, science: britannica.com, Retrieved 6 March, 2020
- Bender, Carl M. (2007-06-01). “Making Sense of Non-Hermitian Hamiltonians”. Reports on Progress in Physics. 70 (6): 947–1018. Arxiv:hep-th/0703096. Bibcode:2007rpph...70..947B. Doi:10.1088/0034-4885/70/6/R03. ISSN 0034-4885
- Wave-function-and-its-physical-significance: winnerscience.com, Retrieved 4 January, 2020
- David Griffiths (2008). Introduction to elementary particles. Wiley-VCH. Pp. 162–. ISBN 978-3-527-40601-2. Retrieved 27 June 2011
- Expect, quantum, hbase: phy-astr.gsu.edu, Retrieved 8 May, 2020

3

One-Dimensional Potentials

One-Dimensional potentials include one-dimensional finite and infinite potential well, one-dimensional potential barrier, one-dimensional step potential, etc. This chapter has been carefully written to provide an easy understanding of the aspects related to one-dimensional potentials.

BOUND STATES

A system of particles is said to be in a bound state if the relative motion of the particles occurs in a bounded region of space (is finite) over a period that is long compared with the periods characteristic of the given system. Nature offers many examples of bound states—from star clusters and macroscopic bodies to such very small entities as atoms, molecules, and atomic nuclei. Many of the particles often referred to as elementary may actually be bound states of other particles.

To form a bound state, attractive forces must exist between at least some particles of the system at certain distances between the particles. For stable bound states, the mass of the system is less than the sum of the masses of the constituent particles. The difference Am determines the binding energy of the system: $E_{binding} = \Delta mc^2$, where c is the speed of light in a vacuum.

Bound states are described in classical mechanics by finite solutions of the equations of motion of the system, where the trajectories of all particles in the system are concentrated in a bounded region of space. An example is Kepler's problem of the motion of a particle or planet in a gravitational field. In classical mechanics a system of two attracting particles can always form a bound state. If the region of distances at which particles are attracted is separated by a potential barrier from the region in which the particles repulse each other, then the particles can also form stable bound states if their motion obeys the laws of classical mechanics.

In quantum mechanics, in contrast to classical mechanics, the formation of a bound state of particles requires that the potential energy of attraction and the range of the forces be sufficiently great. Moreover, in a potential well of the type illustrated in figure, particles can escape from the region of attraction through the tunnel effect. Thus, stable bound states are not formed if the energy of a particle is greater than the potential at infinity. If, however, the probability of tunneling through the barrier is small (at the classical limit it is equal to zero), then a particle can be located in such a potential well for a period that is long compared with the periods of motion in the well. For this reason, in addition to stable bound states there exist unstable bound states, which decay with time; such states are said to be metastable or quasi-stable. The nuclei of some heavy elements, for example, are unstable bound states with respect to alpha decay or fission.

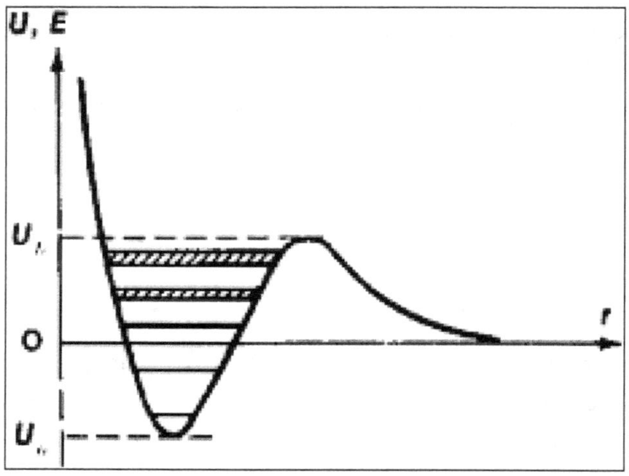

An example of the dependence of the potential energy U on the distance r between particles. This example illustrates the existence of regions of stable and quasi-stable bound states. Stable bound states lie in the region of energies E < O; the energies here are less than the value of U as r. To these states there correspond discrete energy levels. When E > O, stable bound states do not exist. In, however, the region O < E < U_b where U_b is the height of the potential barrier, quasi-stable bound states can exist at certain values of E. The lifetime of such a state is determined by the probability of tunneling through the potential barrier and can be very great, especially for particles of large mass. In the case of macroscopic bodies, whose motion is described by the laws of classical mechanics, stable bound states can have any energy in the region U_o < E < U_b.

In the extremely relativistic case, where the binding energy of the system is comparable to the rest energy of the particles in the system, the solution of the bound-state problem requires the use of quantum field theory. Present-day quantum field theory is not able to provide an exact solution to this problem. Some of the approximate methods being developed permit the consideration of stable and unstable "elementary" particles, including resonance particles, in an identical manner. According to some hypotheses, all hadrons, or all strongly interacting particles, are bound states of more fundamental particles of matter called quarks.

ONE-DIMENSIONAL INFINITE POTENTIAL WELL

In quantum mechanics, the particle in a box model (also known as the infinite potential well or the infinite square well) describes a particle free to move in a small space surrounded by impenetrable barriers. The model is mainly used as a hypothetical example to illustrate the differences between classical and quantum systems. In classical systems, for example, a particle trapped inside a large box can move at any speed within the box and it is no more likely to be found at one position than another. However, when the well becomes very narrow (on the scale of a few nanometers), quantum effects become important. The particle may only occupy certain positive energy levels. Likewise, it can never have zero energy, meaning that the particle can never "sit still". Additionally, it is more likely to be found at certain positions than at others, depending on its energy level. The particle may never be detected at certain positions, known as spatial nodes.

The particle in a box model is one of the very few problems in quantum mechanics which can be solved analytically, without approximations. Due to its simplicity, the model allows insight into quantum effects without the need for complicated mathematics. It serves as a simple illustration of how energy quantizations (energy levels), which are found in more complicated quantum systems such as atoms and molecules, come about. It is one of the first quantum mechanics problems taught in undergraduate physics courses, and it is commonly used as an approximation for more complicated quantum systems.

One-dimensional Solution

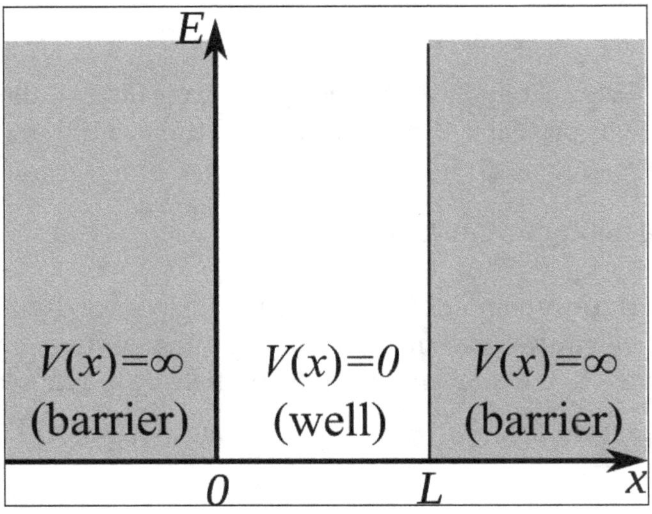

The barriers outside a one-dimensional box have infinitely large potential, while the interior of the box has a constant, zero potential.

The simplest form of the particle in a box model considers a one-dimensional system. Here, the particle may only move backwards and forwards along a straight line with impenetrable barriers at either end. The walls of a one-dimensional box may be visualised as regions of space with an infinitely large potential energy. Conversely, the interior of the box has a constant, zero potential energy. This means that no forces act upon the particle inside the box and it can move freely in that region. However, infinitely large forces repel the particle if it touches the walls of the box, preventing it from escaping. The potential energy in this model is given as:

$$V(x) = \begin{cases} 0, & x_c - \frac{L}{2} < x < x_c + \frac{L}{2}, \\ \infty, & \text{otherwise,} \end{cases}$$

where L is the length of the box, x_c is the location of the center of the box and x is the position of the particle within the box. Simple cases include the centered box ($x_c = 0$) and the shifted box ($x_c = L/2$).

Position Wave Function

In quantum mechanics, the wavefunction gives the most fundamental description of the behavior of a particle; the measurable properties of the particle (such as its position, momentum and energy) may all be derived from the wavefunction. The wavefunction $\psi(x,t)$ can be found by solving the Schrödinger equation for the system:

$$i\hbar \frac{\partial}{\partial t}\psi(x,t) = -\frac{\hbar^2}{2m}\frac{\partial^2}{\partial x^2}\psi(x,t) + V(x)\psi(x,t),$$

where ħ is the reduced Planck constant, m is the mass of the particle, I is the imaginary unit and t is time.

Inside the box, no forces act upon the particle, which means that the part of the wavefunction inside the box oscillates through space and time with the same form as a free particle:

$$(x,t) = [A\sin(kx) + B\cos(kx)]\mathrm{e}^{\ i\ t},$$

where A and B are arbitrary complex numbers. The frequency of the oscillations through space and time is given by the wavenumber k and the angular frequency ω respectively. These are both related to the total energy of the particle by the expression:

$$E = \hbar\omega = \frac{\hbar^2 k^2}{2m},$$

which is known as the dispersion relation for a free particle. Here one must notice that now, since the particle is not entirely free but under the influence of a potential, the energy of the particle given above is not the same thing as $\frac{p^2}{2m}$ where p is the momentum of the particle, and thus the wavenumber k above actually describes the energy states of the particle, not the momentum states (i.e. it turns out that the momentum of the particle is not given by $p = \hbar k$). In this sense, it is quite dangerous to call the number k a wavenumber, since it is not related to momentum like "wavenumber" usually is. The rationale for calling k the wavenumber is that it enumerates the number of crests that the wavefunction has inside the box, and in this sense it is a wavenumber. This discrepancy can be seen more clearly below, when we find out that the energy spectrum of the particle is discrete (only discrete values of energy are allowed) but the momentum spectrum is continuous (momentum can vary continuously) and in particular, the relation $E = \frac{p^2}{2m}$ for the energy and momentum of the particle does not hold. As said above, the reason this relation between energy and momentum does not hold is that the particle is not free, but there is a potential V in the system, and the energy of the particle is $E = T + V$ where T is the kinetic and V the potential energy.

The size (or amplitude) of the wavefunction at a given position is related to the probability of finding a particle there by $P(x,t) = |\psi(x,t)|^2$ The wavefunction must therefore vanish everywhere beyond the edges of the box. Also, the amplitude of the wavefunction may not "jump" abruptly from one point to the next. These two conditions are only satisfied by wavefunctions with the form:

$$\psi_n(x,t) = \begin{cases} A\sin\left(k_n\left(x - x_c + \frac{L}{2}\right)\right)e^{-i\omega_n t}, & x_c - \frac{L}{2} < x < x_c + \frac{L}{2}, \\ 0, & \text{otherwise,} \end{cases}$$

where:

$$k_n = \frac{n\pi}{L},$$

and,

$$E_n = \hbar\omega_n = \frac{n^2\pi^2\hbar^2}{2mL^2},$$

where n is a positive integer (1,2,3,4...). For a shifted box ($x_c = L/2$), the solution is particularly simple. The simplest solutions, $k_n = 0$ or A = 0 both yield the trivial wavefunction $\psi(x) = 0$, which describes a particle that does not exist anywhere in the system. Negative values of n are neglected, since they give wavefunctions identical to the positive n solutions except for a physically unimportant sign change. Here one sees that only a discrete set of energy values and wavenumbers k are allowed for the particle. Usually in quantum mechanics it is also demanded that the derivative of the wavefunction in addition to the wavefunction itself be continuous; here this demand would lead to the only solution being the constant zero function, which is not what we desire, so we give up this demand (as this system with infinite potential can be regarded as a nonphysical abstract limiting case, we can treat it as such and "bend the rules"). Note that giving up this demand means that the wavefunction is not a differentiable function at the boundary of the box, and thus it can be said that the wavefunction does not solve the Schrödinger equation at the boundary points x = 0 and x = L.

Finally, the unknown constant A may be found by normalizing the wavefunction so that the total probability density of finding the particle in the system is 1. It follows that:

$$|A| = \sqrt{\frac{2}{L}}.$$

Thus, A may be any complex number with absolute value $\sqrt{2/L}$; these different values of A yield the same physical state, so $A = \sqrt{2/L}$ can be selected to simplify.

It is expected that the *eigenvalues*, i.e., the energy E_n of the box should be the same regardless of its position in space, but $\psi_n(x,t)$ changes. Notice that $x_c - \frac{L}{2}L$ represents a phase shift in the wave function, This phase shift has no effect when solving the Schrödinger equation, and therefore does not affect the *eigenvalue*.

If we defined the well as centred on the origin, we can rewrite the spacial part of the wave function succinctly as:

$$\psi_n(x) = \begin{cases} \sqrt{\frac{2}{L}}\sin(k_n x) & \text{for } n \text{ even} \\ \sqrt{\frac{2}{L}}\cos(k_n x) & \text{for } n \text{ odd} \end{cases}$$

Momentum Wave Function

The momentum wavefunction is proportional to the Fourier transform of the position wavefunction. With $k = p/\hbar$ (note that the parameter k describing the momentum wavefunction below is not exactly the special k_n above, linked to the energy eigenvalues), the momentum wavefunction is given by:

$$\phi_n(p,t) = \frac{1}{\sqrt{2\pi\hbar}}\int_{-\infty}^{\infty}\Psi_n(x,t)e^{-ikx}\,dx = \sqrt{\frac{L}{\pi\hbar}}\left(\frac{n\pi}{n\pi + kL}\right)\operatorname{sin}c\left(\tfrac{1}{2}(n\pi - kL)\right)e^{-ikx_c}e^{i(n-1)\frac{\pi}{2}}e^{-i\omega_n t},$$

where sinc is the cardinal sine sinc function, sinc(x)=sin(x)/x. For the centered box

($x_c = o$), the solution is real and particularly simple, since the phase factor on the right reduces to unity. (With care, it can be written as an even function of p).

It can be seen that the momentum spectrum in this wave packet is continuous, and one may conclude that for the energy state described by the wavenumber k_n, the momentum can, when measured, also attain *other values* beyond $p = \pm\hbar k_n$.

Hence, it also appears that, since the energy is $E_n = \frac{\hbar^2 k_n^2}{2m}$ for the nth eigenstate, the relation $E = \frac{p^2}{2m}$ does not strictly hold for the measured momentum p; the energy eigenstate is not a momentum eigenstate, and, in fact, not even a superposition of two momentum eigenstates, as one might be tempted to imagine from equation above: peculiarly, it has no well-defined momentum before measurement. $E = \frac{p^2}{2m}$ does not strictly hold for the measured momentum p; the energy eigenstate ψ_n is not a momentum eigenstate, and, in fact, not even a superposition of two momentum eigenstates, as one might be tempted to imagine from equation $\psi(x,t) = [A\sin(kx) + B\cos(kx)]e^{-i\omega t}$, peculiarly, it has no well-defined momentum before measurement.

Position and Momentum Probability Distributions

In classical physics, the particle can be detected anywhere in the box with equal probability. In quantum mechanics, however, the probability density for finding a particle at a given position is derived from the wavefunction as $P(x) = |\psi(x)|^2$. For the particle in a box, the probability density for finding the particle at a given position depends upon its state, and is given by:

$$P_n(x,t) = \begin{cases} \frac{2}{L}\sin^2(k_n(x - x_c + \frac{L}{2})), & x_c - L/2 < x < x_c + L/2, \\ 0, & \text{otherwise,} \end{cases}$$

Thus, for any value of n greater than one, there are regions within the box for which $P(x) = 0$, indicating that *spatial nodes* exist at which the particle cannot be found.

In quantum mechanics, the average, or expectation value of the position of a particle is given by:

$$\langle x \rangle = \int_{-\infty}^{\infty} x P_n(x)\, dx.$$

or the steady state particle in a box, it can be shown that the average position is always $\langle x \rangle = x_c$, regardless of the state of the particle. For a superposition of states, the expectation value of the position will change based on the cross term which is proportional to $\cos(\omega t)$.

The variance in the position is a measure of the uncertainty in position of the particle:

$$\mathrm{Var}(x) = \int_{-\infty}^{\infty} (x - \langle x \rangle)^2 P_n(x)\, dx = \frac{L^2}{12}\left(1 - \frac{6}{n^2\pi^2}\right)$$

The probability density for finding a particle with a given momentum is derived from the wavefunction as $P(x) = |\phi(x)|^2$ As with position, the probability density for finding the particle at a given momentum depends upon its state, and is given by:

$$P_n(p) = \frac{L}{\pi\hbar}\left(\frac{n\pi}{n\pi + kL}\right)^2 \mathrm{sinc}^2\left(\tfrac{1}{2}(n\pi - kL)\right)$$

where, again, $k = p/\hbar$ The expectation value for the momentum is then calculated to be zero, and the variance in the momentum is calculated to be:

$$\mathrm{Var}(p) = \left(\frac{\hbar n\pi}{L}\right)^2$$

The uncertainties in position and momentum (Δx and Δp) are defined as being equal to the square root of their respective variances, so that:

$$\Delta x \Delta p = \frac{\hbar}{2}\sqrt{\frac{n^2\pi^2}{3} - 2}$$

This product increases with increasing *n*, having a minimum value for *n=1*. The value of this product for *n=1* is about equal to 0.568 $\hbar$ which obeys the Heisenberg uncertainty principle, which states that the product will be greater than or equal to $\hbar/2$.

Another measure of uncertainty in position is the information entropy of the probability distribution H_x:

$$H_x = \int_{-\infty}^{\infty} P_n(x)\log(P_n(x)x_0)\, dx = \log\left(\frac{2L}{e\, x_0}\right)$$

where x_0 is an arbitrary reference length.

Another measure of uncertainty in momentum is the information entropy of the probability distribution H_p:

$$H_p(n) = \int_{-\infty}^{\infty} P_n(p)\log(P_n(p)p_0)\, dp$$

$$\lim_{n\to\infty} H_p(n) = \log\left(\frac{4\pi\hbar\, e^{2(1-\gamma)}}{L\, p_0}\right)$$

where γ is Euler's constant. The quantum mechanical entropic uncertainty principle states that for $x_0\, p_0 = \hbar$.

$$H_x + H_p(n) \geq \log(e\pi) \approx 214473 \text{ (nats)}.$$

For $x_0\, p_0 = \hbar$ the sum of the position and momentum entropies yields:

$$H_x + H_p(\infty) = \log(8\pi\, e^{1-2\gamma}) \approx 306974 \text{ (nats)},$$

which satisfies the quantum entropic uncertainty principle.

Energy Levels

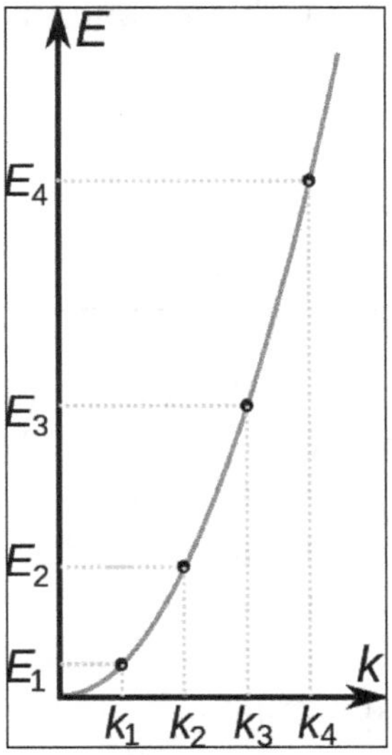

The energy of a particle in a box (black circles) and a free particle (grey line) both depend upon wavenumber in the same way. However, the particle in a box may only have certain, discrete energy levels.

The energies which correspond with each of the permitted wavenumbers may be written as:

$$E_n = \frac{n^2\hbar^2\pi^2}{2mL^2} = \frac{n^2h^2}{8mL^2}$$

The energy levels increase with n^2, meaning that high energy levels are separated from each other by a greater amount than low energy levels are. The lowest possible energy for the particle (its *zero-point energy*) is found in state 1, which is given by:

$$E_1 = \frac{\hbar^2\pi^2}{2mL^2}.$$

The particle, therefore, always has a positive energy. This contrasts with classical systems, where the particle can have zero energy by resting motionlessly. This can be explained in terms of the uncertainty principle, which states that the product of the uncertainties in the position and momentum of a particle is limited by:

$$\Delta x \Delta p \geq \frac{\hbar}{2}$$

It can be shown that the uncertainty in the position of the particle is proportional to the width of the box. Thus, the uncertainty in momentum is roughly inversely proportional to the width of the box. The kinetic energy of a particle is given by $E = p^2/(2m)$ and hence the minimum kinetic energy of the particle in a box is inversely proportional to the mass and the square of the well width, in qualitative agreement with the calculation above.

ONE-DIMENSIONAL FINITE POTENTIAL WELL

The finite potential well (also known as the finite square well) is a concept from quantum mechanics. It is an extension of the infinite potential well, in which a particle is confined to a box, but one which has finite potential walls. Unlike the infinite potential well, there is a probability associated with the particle being found outside the box. The quantum mechanical interpretation is unlike the classical interpretation, where if the total energy of the particle is less than the potential energy barrier of the walls it cannot be found outside the box. In the quantum interpretation, there is a non-zero probability of the particle being outside the box even when the energy of the particle is less than the potential energy barrier of the walls.

Particle in a 1-dimensional Box

For the 1-dimensional case on the x-axis, the time-independent Schrödinger equation can be written as:

$$-\frac{\hbar^2}{2m}\frac{d^2\psi}{dx^2} + V(x)\psi = E\psi$$

where:

$$\hbar = \frac{h}{2\pi},$$

- h is Planck's constant,
- m is the mass of the particle,

- ψ is the (complex valued) wavefunction that we want to find,
- $V(x)$ is a function describing the potential energy at each point x,
- E is the energy, a real number, sometimes called eigenenergy.

For the case of the particle in a 1-dimensional box of length L, the potential is V_0 outside the box, and zero for x between $-L/2$ and $L/2$. The wavefunction is considered to be made up of different wavefunctions at different ranges of x, depending on whether x is inside or outside of the box. Therefore, the wavefunction is defined such that:

$$\psi = \begin{cases} \psi_1, & \text{if } x < -L/2 \text{ (the region outside the box)} \\ \psi_2, & \text{if } -L/2 < x < L/2 \text{ (the region inside the box)} \\ \psi_3 & \text{if } x > L/2 \text{ (the region outside the box)} \end{cases}$$

Inside the Box

For the region Inside the box $V(x) = 0$ and equation $-\frac{\hbar^2}{2m}\frac{d^2\psi}{dx^2} + V(x)\psi = E\psi$ reduces to:

$$-\frac{\hbar^2}{2m}\frac{d^2\psi_2}{dx^2} = E\psi_2.$$

Letting,

$$k = \frac{\sqrt{2mE}}{\hbar},$$

the equation becomes,

$$\frac{d^2\psi_2}{dx^2} = -k^2\psi_2.$$

This is a well-studied differential equation and eigenvalue problem with a general solution of:

$$\psi_2 = A\sin(kx) + B\cos(kx)$$

hence,

$$E = \frac{k^2\hbar^2}{2m}.$$

Here, A and B can be any complex numbers, and k can be any real number.

Outside the Box

For the region Outside of the box, since the potential is constant, $V(x) = V_0$ and equation $-\frac{\hbar^2}{2m}\frac{d^2\psi}{dx^2} + V(x)\psi = E\psi$ becomes:

$$-\frac{\hbar^2}{2m}\frac{d^2\psi_1}{dx^2} = (E - V_0)\psi_1$$

There are two possible families of solutions, depending on whether E is less than V_0 (the particle is bound in the potential) or E is greater than V_0 (the particle is free).

For a free particle, $E > V_0$, and letting,

$$k' = \frac{\sqrt{2m(E - V_0)}}{\hbar}$$

produces,

$$\frac{d^2\psi_1}{dx^2} = -k'^2\psi_1$$

with the same solution form as the inside-well case:

$$\psi_1 = C\sin(k'x) + D\cos(k'x)$$

This analysis will focus on the bound state, where $V_0 > E$. Letting,

$$\alpha = \frac{\sqrt{2m(V_0 - E)}}{\hbar}$$

Produces,

$$\frac{d^2\psi_1}{dx^2} = \alpha^2\psi_1$$

where the general solution is exponential:

$$\psi_1 = Fe^{-\alpha x} + Ge^{\alpha x}$$

Similarly, for the other region outside the box:

$$\psi_3 = He^{-\alpha x} + Ie^{\alpha x}$$

Now in order to find the specific solution for the problem at hand, we must specify the appropriate boundary conditions and find the values for A, B, F, G, H and I that satisfy those conditions.

Finding Wavefunctions for the Bound State

Solutions to the Schrödinger equation must be continuous, and continuously differentiable. These requirements are boundary conditions on the differential equations previously derived, that is, the matching conditions between the solutions inside and outside the well.

In this case, the finite potential well is symmetrical, so symmetry can be exploited to reduce the necessary calculations.

$$\psi = \begin{cases} \psi_1, & \text{if } x < -L/2 \text{ (the region outside the box)} \\ \psi_2, & \text{if } -L/2 < x < L/2 \text{ (the region inside the box)} \\ \psi_3 & \text{if } x > L/2 \text{ (the region outside the box)} \end{cases}$$

where we found ψ_1, ψ_2 and ψ_3 to be:

$$\psi_1 = Fe^{-\alpha x} + Ge^{\alpha x}$$
$$\psi_2 = A\sin(kx) + B\cos(kx)$$
$$\psi_3 = He^{-\alpha x} + Ie^{\alpha x}$$

We see that as x goes to $-\infty$, the F term goes to infinity. Likewise, as x goes to $+\infty$, the I term goes to infinity. In order for the wave function to be square integrable, we must set $F = I = 0$, and we have:

$\psi_1 = Ge^{\alpha x}$ and $\psi_3 = He^{-\alpha x}$.

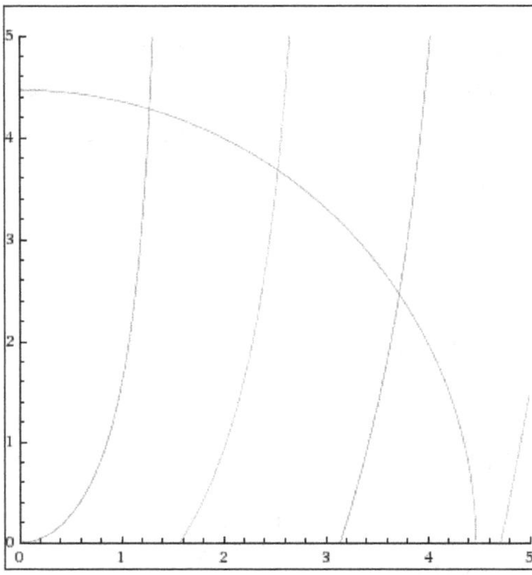

Next, we know that the overall ψ function must be continuous and differentiable. In other words, the values of the functions and their derivatives must match up at the dividing points:

$$\psi_1(-L/2) = \psi_2(-L/2) \qquad \psi_2(L/2) = \psi_3(L/2)$$
$$\frac{d\psi_1}{dx}(-L/2) = \frac{d\psi_2}{dx}(-L/2) \qquad \frac{d\psi_2}{dx}(L/2) = \frac{d\psi_3}{dx}(L/2)$$

These equations have two sorts of solutions, symmetric, for which A = o and G = H, and antisymmetric, for which B = o and G = – H. For the symmetric case we get:

$$He^{-\alpha L/2} = B\cos(kL/2)$$
$$-\alpha He^{-\alpha L/2} = -kB\sin(kL/2)$$

so taking the ratio gives,

$$\alpha = k\tan(kL/2).$$

Similarly for the antisymmetric case we get:

$$\alpha = -k\cot(kL/2).$$

Recall that both α and k.

Depend on the energy. What we have found is that the continuity conditions cannot be satisfied for an arbitrary value of the energy. Only certain energy values, which are solutions to one or either of these two equations, are allowed. Hence we find that the energy levels of the system below V_0 are discrete; the corresponding eigenfunctions are bound states. (By contrast, for the energy levels above V_0 are continuous).

The energy equations cannot be solved analytically. Nevertheless, we will see that in the symmetric case, there always exists at least one bound state, even if the well is very shallow. Graphical or numerical solutions to the energy equations are aided by rewriting them a little. If we introduce the dimensionless variables $u = \alpha L/2$ and $v = kL/2$, and note from the definitions of α and k that $u^2 = u_0^2 - v^2$, where $u_0^2 = mL^2V_0/2\hbar^2$, the master equations read:

$$\sqrt{u_0^2 - v^2} = \begin{cases} v\tan v, & \text{(symmetric case)} \\ -v\cot v, & \text{(antisymmetric case)} \end{cases}$$

In the plot to the right, for $u_0^2 = 20$, solutions exist where the blue semicircle intersects the purple or grey curves ($v\tan v$ and $-v\cot v$). Each purple or grey curve represents a possible solution, v_i within the range $\frac{\pi}{2}(i-1) \le v_i < \frac{\pi}{2}i$. The total number of solutions, N(i.e., the number of purple/grey curves that are intersected by the blue circle) is therefore determined by dividing the radius of the blue circle, u_0 by the range of each solution $\pi/2$.

$$N = \left\lfloor \frac{2u_0}{\pi} \right\rfloor + 1 = \left\lceil \frac{2u_0}{\pi} \right\rceil$$

In this case there are exactly three solutions, since:

$$N = \left\lfloor 2\sqrt{20}/\pi \right\rfloor + 1 = \lfloor 2.85 \rfloor + 1 = 2 + 1 = 3.$$

$v_1 = 1.28, v_2 = 2.54$ *and* $v_3 = 3.73$, with the corresponding energies:

$$E_n = \frac{2\hbar^2 v_n^2}{mL^2}.$$

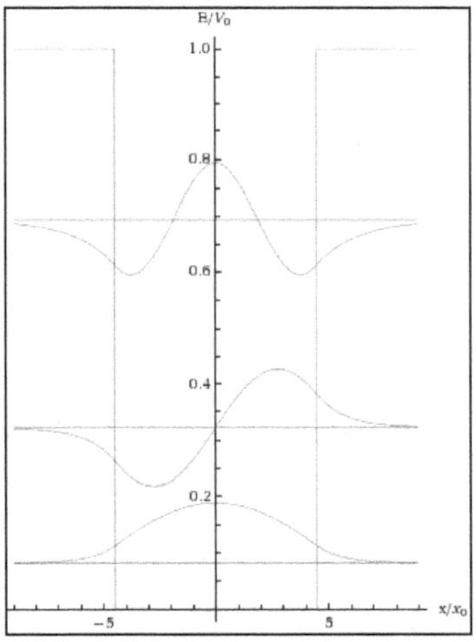

If we want, we can go back and find the values of the constants A, B, G, H in the equations now (we also need to impose the normalisation condition). On the right we show the energy levels and wave functions in this case (where $x_0 \equiv \hbar / \sqrt{2mV_0}$).

We note that however small u_0 is (however shallow or narrow the well), there is always at least one bound state.

Two special cases are worth noting. As the height of the potential becomes large, $V_0 \to \infty$, the radius of the semicircle gets larger and the roots get closer and closer to the values $v_n = n\pi/2$, and we recover the case of the infinite square well.

The other case is that of a very narrow, deep well - specifically the case $V_0 \to \infty$ and $L \to 0$ with $V_0 L$ fixed. As $u_0 \propto V_0 L^2$ it will tend to zero, and so there will only be one bound state. The approximate solution is then $v^2 = u_0^2 - u_0^4$, and the energy tends to $E = -mL^2V_0^2 / 2\hbar^2$. But this is just the energy of the bound state of a Delta function potential of strength $V L$, as it should be.

A simpler graphical solution for the energy levels can be obtained by normalizing the potential and the energy through multiplication by $8mL^2 / h^2$. The normalized quantities are:

$$\tilde{V}_0 = V_0 \frac{8m}{h^2} L^2 \qquad \tilde{E} = E \frac{8m}{h^2} L^2$$

giving directly the relation between the allowed couples (V_0, E) as:

$$\sqrt{\tilde{V}_0} = \sqrt{\tilde{E}} \left| \sec(\sqrt{\tilde{E}}\, \pi / 2) \right|, \qquad \sqrt{\tilde{V}_0} = \sqrt{\tilde{E}} \left| \csc(\sqrt{\tilde{E}}\, \pi / 2) \right|$$

for the even and odd parity wave functions, respectively. In the previous equations only the positive derivative parts of the functions have to be considered. The chart giving directly the allowed couples (V_0, E) is reported in the figure.

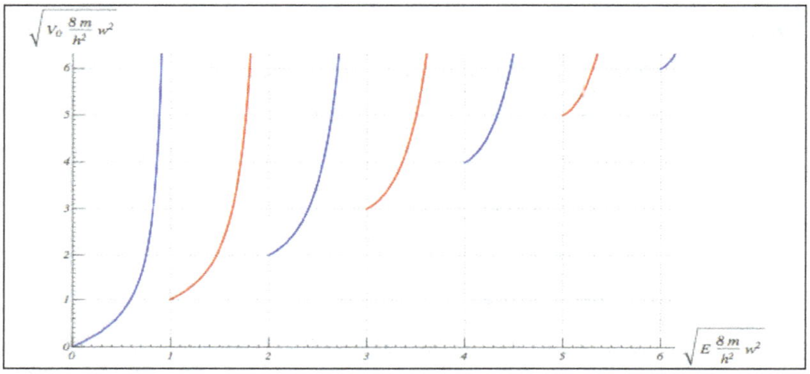

Particle in a quantum finite well: possible values of (V_0, E) for a particle of mass m and energy $E \leq V_0$ in a finite potential well of depth V_0 and width w.
Blue: even parity solutions. Red: odd parity solutions.

Unbound States

If we solve the time-independent Schrödinger equation for an energy $E > V_0$, the solutions will be oscillatory both inside and outside the well. Thus, the solution is never square integrable; that is, it is always a non-normalizable state. This does not mean, however, that it is impossible for a quantum particle to have energy greater than V_0, it merely means that the system has continuous spectrum above V_0. The non-normalizable eigenstates are close enough to being square integrable that they still contribute to the spectrum of the Hamiltonian as an unbounded operator.

Asymmetrical Well

Consider a one-dimensional asymmtrical potential well given by the potential:

$$V(x) = \begin{cases} V_1, & \text{if } -\infty < x < 0 \text{ (the region outside the box)} \\ 0, & \text{if } 0 < x < a \text{ (the region inside the box)} \\ V_2 & \text{if } a < x < \infty \text{ (the region outside the box)} \end{cases}$$

with $V_2 > V_1$. The corresponding solution for the wave function with $E < V_1$ is found to be:

$$\psi(x) = \begin{cases} c_1 e^{k_1 x}, & \text{for } x < 0, \text{where } k_1 = \sqrt{(2m/\hbar^2)(V_1 - E)} \\ c\sin(kx + \delta), & \text{for } 0 < x < a, \text{where } k = \sqrt{2mE/\hbar^2} \\ c_2 e^{-k_2 x}, & \text{for } x > a, \text{where } k_2 = \sqrt{(2m/\hbar^2)(V_2 - E)} \end{cases}$$

and,

$$\sin\delta = \frac{k\hbar}{\sqrt{2mV_1}}.$$

The energy levels $E = k^2\hbar^2/(2m)$ is determined once k is solved as a root of the following transcendental equation:

$$ka = n\pi - \sin^{-1}\left[k\hbar/\sqrt{2mV_1}\right] - \sin^{-1}\left[k\hbar/\sqrt{2mV_2}\right]$$

where n = 1, 2, 3, ... Existence of root to above equation is not always guaranteed, for example, one can always find a value of a so small, that for given values of V_1 and V_2, there exists no discrete energy level. The results of symmetrical well is obtained from above equation by setting $V_1 = V_2 = V_o$.

ONE-DIMENSIONAL POTENTIAL BARRIER

In quantum mechanics, the rectangular (or, at times, square) potential barrier is a standard one-dimensional problem that demonstrates the phenomena of wave-mechanical tunneling (also called "quantum tunneling") and wave-mechanical reflection. The problem consists of solving the one-dimensional time-independent Schrödinger equation for a particle encountering a rectangular potential energy barrier. It is usually assumed, as here, that a free particle impinges on the barrier from the left.

Although classically a particle behaving as a point mass would be reflected, a particle actually behaving as a matter wave has a non-zero probability of penetrating the barrier and continuing its travel as a wave on the other side. In classical wave-physics, this effect is known as evanescent wave coupling. The likelihood that the particle will pass through the barrier is given by the transmission coefficient, whereas the likelihood that it is reflected is given by the reflection coefficient. Schrödinger's wave-equation allows these coefficients to be calculated.

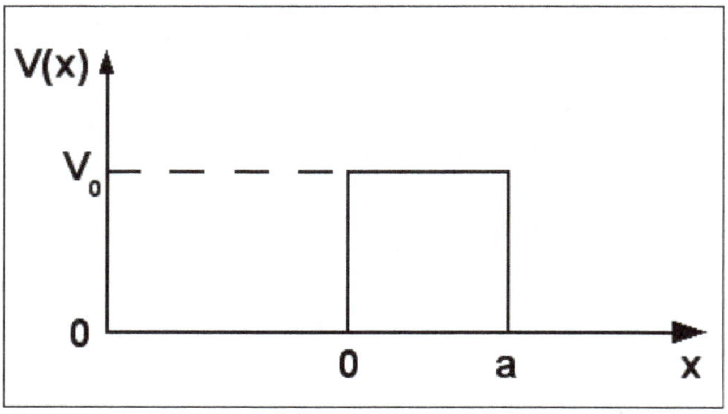

Calculation

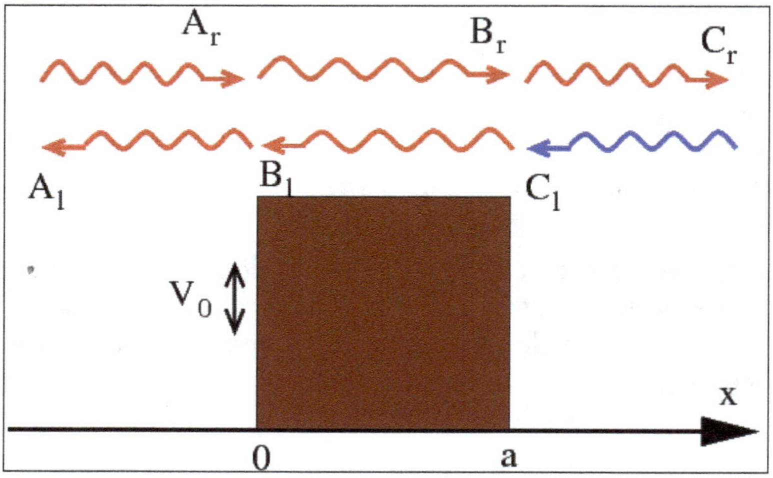

Scattering at a finite potential barrier of height V_0. The amplitudes and direction of left and right moving waves are indicated. In red, those waves used for the derivation of the reflection and transmission amplitude. $E > V_0$ for this illustration.

The time-independent Schrödinger equation for the wave function $\psi(x)$ reads:

$$H\psi(x) = \left[-\frac{\hbar^2}{2m}\frac{d^2}{dx^2} + V(x)\right]\psi(x) = E\psi(x)$$

where H is the Hamiltonian, ħ is the (reduced) Planck constant, m is the mass, E the energy of the particle and,

$$V(x) = V_0[\Theta(x) - \Theta(x-a)]$$

is the barrier potential with height $V_0 > 0$ and width a.

$$\Theta(x) = 0, x < 0; \Theta(x) = 1, x > 0$$

is the Heaviside step function, i.e.

$$V(x) = \begin{cases} 0 & \text{if } x < 0 \\ V_0 & \text{if } 0 < x < a \\ 0 & \text{if } a < x \end{cases}$$

The barrier is positioned between x = o and x = a. The barrier can be shifted to any x position without changing the results. The first term in the, Hamiltonian, $-\frac{\hbar^2}{2m}\frac{d^2}{dx^2}\psi$ is the kinetic energy.

The barrier divides the space in three parts ($x < 0, 0 < x < a, x > a$ In any of these parts, the potential is constant, meaning that the particle is quasi-free, and the solution of the Schrödinger equation can be written as a superposition of left and right moving waves. If E > V_o,

$$\begin{aligned}
\psi_L(x) &= A_r e^{ik_0x} + A_l e^{-ik_0x} & x < 0 \\
\psi_C(x) &= B_r e^{ik_1x} + B_l e^{-ik_1x} & 0 < x < a \\
\psi_R(x) &= C_r e^{ik_0x} + C_l e^{-ik_0x} & x > a
\end{aligned}$$

where the wave numbers are related to the energy via:

$$\begin{aligned}
k_0 &= \sqrt{2mE / \hbar^2} & x < 0 \quad or \quad x > a \\
k_1 &= \sqrt{2m(E - V_0) / \hbar^2} & 0 < x < a.
\end{aligned}$$

The index r/l on the coefficients A and B denotes the direction of the velocity vector. Note that, if the energy of the particle is below the barrier height, k_1 becomes imaginary and the wave function is exponentially decaying within the barrier. Nevertheless, we keep the notation r/l even though the waves are not propagating anymore in this case. Here we assumed E ≠ V_o. The case E = V_o is treated below.

The coefficients A, B, C have to be found from the boundary conditions of the wave function at x = o and x = a. The wave function and its derivative have to be continuous everywhere, so:

$$\begin{aligned}
&\psi_L(0) = \psi_C(0) \\
&\frac{d}{dx}\psi_L(0) = \frac{d}{dx}\psi_C(0) \\
&\psi_C(a) = \psi_R(a) \\
&\frac{d}{dx}\psi_C(a) = \frac{d}{dx}\psi_R(a).
\end{aligned}$$

Inserting the wave functions, the boundary conditions give the following restrictions on the coefficients:

$$A_r + A_l = B_r + B_l$$
$$ik_0(A_r - A_l) = ik_1(B_r - B_l)$$
$$B_r e^{iak_1} + B_l e^{-iak_1} = C_r e^{iak_0} + C_l e^{-iak_0}$$
$$ik_1(B_r e^{iak_1} - B_l e^{-iak_1}) = ik_0(C_r e^{iak_0} - C_l e^{-iak_0}).$$

$E = V_0$

If the energy equals the barrier height, the second differential of the wavefunction inside the barrier region is 0, and hence the solutions of the Schrödinger equation are not exponentials anymore but linear functions of the space coordinate,

$$\psi_C(x) = B_1 + B_2 x \quad 0 < x < a.$$

The complete solution of the Schrödinger equation is found in the same way as above by matching wave functions and their derivatives at x = 0 and x = a. That results in the following restrictions on the coefficients:

$$A_r + A_l = B_1$$
$$ik_0(A_r - A_l) = B_2$$
$$B_1 + B_2 a = C_r e^{iak_0} + C_l e^{-iak_0}$$
$$B_2 = ik_0(C_r e^{iak_0} - C_l e^{-iak_0}).$$

Transmission and Reflection

At this point, it is instructive to compare the situation to the classical case. In both cases, the particle behaves as a free particle outside of the barrier region. A classical particle with energy E larger than the barrier height V_0 would *always* pass the barrier, and a classical particle with $E < V_0$ incident on the barrier would *always* get reflected.

To study the quantum case, consider the following situation: a particle incident on the barrier from the left side (A_r). It may be reflected (A_l) or transmitted (C_r).

To find the amplitudes for reflection and transmission for incidence from the left, we put in the above equations $A_r = 1$ (incoming particle), $A_l = r$ (reflection), $C_l = 0$(no incoming particle from the right), and $C_r = t$ (transmission). We then eliminate the coefficients B_l, B_r from the equation and solve for r and t.

The result is:

$$t = \frac{4k_0 k_1 e^{-ia(k_0 - k_1)}}{(k_0 + k_1)^2 - e^{2iak_1}(k_0 - k_1)^2}$$

$$r = \frac{(k_0^2 - k_1^2)\sin(ak_1)}{2ik_0k_1\cos(ak_1) + (k_0^2 + k_1^2)\sin(ak_1)}.$$

Due to the mirror symmetry of the model, the amplitudes for incidence from the right are the same as those from the left. Note that these expressions hold for any energy $E > 0$.

Analysis of the Obtained Expressions

Expressions $E < V_o$

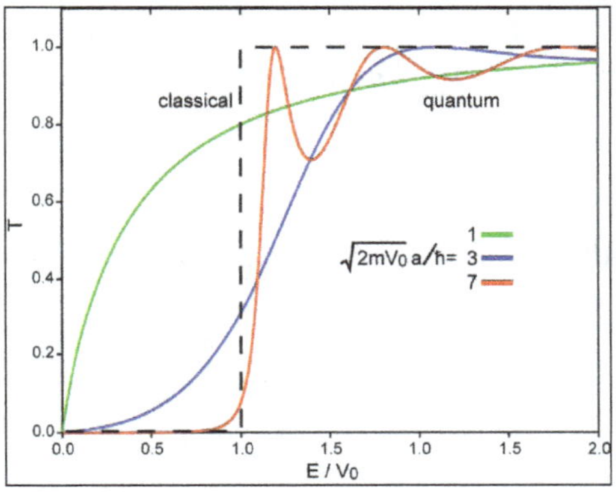

Transmission probability of a finite potential barrier for $\sqrt{2mV_0}a/\hbar$ =1, 3, and 7. Dashed: classical result. Solid line: quantum mechanics.

The surprising result is that for energies less than the barrier height, $E < V_0$ there is a non-zero probability:

$$T = |t|^2 = \frac{1}{1 + \dfrac{V_0^2 \sinh^2(k_1 a)}{4E(V_0 - E)}}$$

for the particle to be transmitted through the barrier, with $k_1 = \sqrt{2m(V_0 - E)/\hbar^2}$. This effect, which differs from the classical case, is called quantum tunneling. The transmission is exponentially suppressed with the barrier width, which can be understood from the functional form of the wave function: Outside of the barrier it oscillates with wave vector k_0, whereas within the barrier it is exponentially damped over a distance $1/k_1$. If the barrier is much wider than this decay length, the left and right part are virtually independent and tunneling as a consequence is suppressed.

Expressions $E > V_0$

In this case,

$$T = |t|^2 = \frac{1}{1 + \dfrac{V_0^2 \sin^2(k_1 a)}{4E(E - V_0)}},$$

where $k_1 = \sqrt{2m(E - V_0) / \hbar^2}$.

Equally surprising is that for energies larger than the barrier height, E > V_o the particle may be reflected from the barrier with a non-zero probability:

$$R = |r|^2 = 1 - T.$$

The transmission and reflection probabilities are in fact oscillating with $k_1 a$. The classical result of perfect transmission without any reflection $(T = 1, R = 0)$ is reproduced not only in the limit of high energy $E \gg V_0$ but also when the energy and barrier width satisfy $k_1 a = n\pi$, where $n = 0, 1, 2,.$ $(E / V_0 = 1.2)$. Note that the probabilities and amplitudes as written are for any energy the barrier height.

Expressions $E = V_o$

The transmission probability at $E = V_0$ evaluates to:

$$T = \frac{1}{1 + ma^2 V_0 / 2\hbar^2}.$$

Remarks and Applications

The calculation presented above may at first seem unrealistic and hardly useful. However it has proved to be a suitable model for a variety of real-life systems. One such example are interfaces between two conducting materials. In the bulk of the materials, the motion of the electrons is quasi-free and can be described by the kinetic term in the above Hamiltonian with an effective mass m. Often the surfaces of such materials are covered with oxide layers or are not ideal for other reasons. This thin, non-conducting layer may then be modeled by a barrier potential as above. Electrons may then tunnel from one material to the other giving rise to a current.

The operation of a scanning tunneling microscope (STM) relies on this tunneling effect. In that case, the barrier is due to the gap between the tip of the STM and the underlying object. Since the tunnel current depends exponentially on the barrier width, this device is extremely sensitive to height variations on the examined sample.

The above model is one-dimensional, while space is three-dimensional. One should solve the Schrödinger equation in three dimensions. On the other hand, many systems only change along one coordinate direction and are translationally invariant along the others; they are separable. The Schrödinger equation may then be reduced to the case considered here by an ansatz for the wave function of the type:

$$\Psi(x, y, z) = \psi(x)\phi(y, z).$$

ONE-DIMENSIONAL STEP POTENTIAL

In quantum mechanics and scattering theory, the one-dimensional step potential is an idealized system used to model incident, reflected and transmitted matter waves. The problem consists of solving the time-independent Schrödinger equation for a particle with a step-like potential in one dimension. Typically, the potential is modelled as a Heaviside step function.

Calculation

Schrödinger Equation and Potential Function

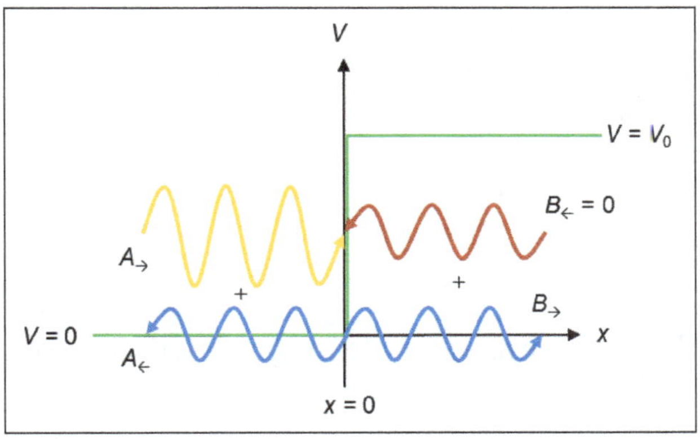

Scattering at a finite potential step of height V_0, shown in green. The amplitudes and direction of left and right moving waves are indicated. Yellow is the incident wave, blue are reflected and transmitted, red does not occur.

The time-independent Schrödinger equation for the wave function $\psi(x)$ is:

$$H\psi(x) = \left[-\frac{\hbar^2}{2m}\frac{d^2}{dx^2} + V(x) \right]\psi(x) = E\psi(x),$$

where H is the Hamiltonian, ħ is the reduced Planck constant, m is the mass, E the energy of the particle. The step potential is simply the product of V_0, the height of the barrier, and the Heaviside step function:

$$V(x) = \begin{cases} 0, & x < 0 \\ V_0, & x \geq 0 \end{cases}$$

The barrier is positioned at x = 0, though any position x_0 may be chosen without changing the results, simply by shifting position of the step by $-x_0$.

The first term in the Hamiltonian, $-\frac{\hbar^2}{2m}\frac{d^2}{dx^2}\psi$ is the kinetic energy of the particle.

Solution:

The step divides space in two parts: x < 0 and x > 0. In any of these parts the potential is constant, meaning the particle is quasi-free, and the solution of the Schrödinger equation can be written as a superposition of left and right moving waves:

$$\psi_1(x) = \frac{1}{\sqrt{k_1}}\left(A_{\rightarrow}e^{ik_1x} + A_{\leftarrow}e^{-ik_1x}\right) \quad x < 0,$$

$$\psi_2(x) = \frac{1}{\sqrt{k_2}}\left(B_{\rightarrow}e^{ik_2x} + B_{\leftarrow}e^{-ik_2x}\right) \quad x > 0$$

where subscripts 1 and 2 denote the regions x < 0 and x > 0 respectively, the subscripts (→) and (←) on the amplitudes A and B denote the direction of the particle's velocity vector: right and left respectively.

The coefficients $1/\sqrt{k_1}$ and $1/\sqrt{k_2}$ are normalization constants. The wave vectors in the respective regions being:

$$k_1 = \sqrt{2mE/\hbar^2},$$

$$k_2 = \sqrt{2m(E - V_0)/\hbar^2}$$

both of which have the same form as the De Broglie relation (in one dimension):

$$p = \hbar k.$$

Boundary Conditions

The coefficients A, B have to be found from the boundary conditions of the wave function at x = 0. The wave function and its derivative have to be continuous everywhere, so:

$$\psi_1(0) = \psi_2(0),$$

$$\frac{d}{dx}\psi_1(x)\Big|_{x=0} = \frac{d}{dx}\psi_2(x)\Big|_{x=0}.$$

Inserting the wave functions, the boundary conditions give the following restrictions on the coefficients:

$$(A_{\rightarrow} + A_{\leftarrow}) = (B_{\rightarrow} + B_{\leftarrow})$$

$$k_1(A_{\rightarrow} - A_{\leftarrow}) = k_2(B_{\rightarrow} - B_{\leftarrow})$$

Transmission and Reflection

It is useful to compare the situation to the classical case. In both cases, the particle behaves as a free particle outside of the barrier region. A classical particle with energy E larger than the barrier height V_0 will be slowed down but never reflected by the barrier, while a classical particle with $E < V_0$ incident on the barrier from the left would always be reflected. Once we have found the quantum-mechanical result we will return to the question of how to recover the classical limit.

To study the quantum case, consider the following situation: a particle incident on the barrier from the left side $A_{\rightarrow}$. It may be reflected ($A_{\leftarrow}$) or transmitted ($B_{\rightarrow}$). Here and in the following assume $E > V_0$.

To find the amplitudes for reflection and transmission for incidence from the left, we set in the above equations $A_{\rightarrow} = 1$ (incoming particle), $A_{\leftarrow} = \sqrt{R}$ (reflection), $B_{\leftarrow} = 0$ (no incoming particle from the right) and $B_{\rightarrow} = \sqrt{T k_1/k_2}$ (transmission). We then solve for T and R.

The result is:

$$\sqrt{T} = \frac{2\sqrt{k_1 k_2}}{k_1 + k_2}$$

$$\sqrt{R} = \frac{k_1 - k_2}{k_1 + k_2}.$$

The model is symmetric with respect to a parity transformation and at the same time interchange k_1 and k_2. For incidence from the right we have therefore the amplitudes for transmission and reflection:

$$\sqrt{T'} = \sqrt{T} = \frac{2\sqrt{k_1 k_2}}{k_1 + k_2}$$

$$\sqrt{R'} = -\sqrt{R} = \frac{k_2 - k_1}{k_1 + k_2}.$$

Analysis of the Expressions

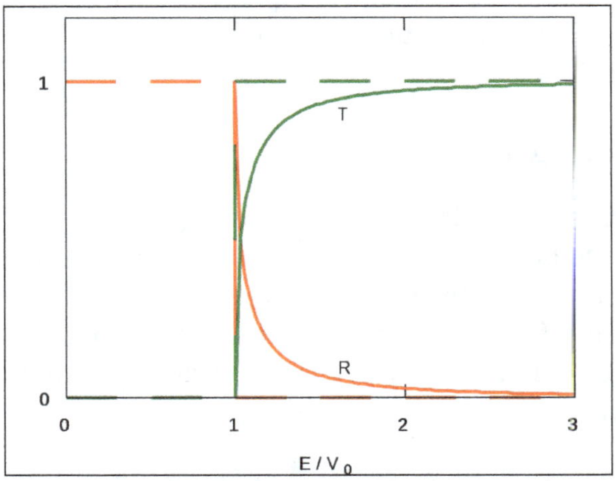

Reflection and transmission probability at a Heaviside-step potential. Dashed: classical result. Solid lines: quantum mechanics. For $E < V_0$ the classical and quantum problem give the same result.

Energy Less than Step Height ($E < V_0$)

For energies $E < V_0$, the wave function to the right of the step is exponentially decaying over a distance $1/(k_2)$.

Energy Greater than Step Height ($E > V_0$)

In this energy range the transmission and reflection coefficient differ from the classical case. They are the same for incidence from the left and right:

$$T = |T'| = \frac{4k_1 * k_2}{(k_1 + k_2)^2}$$

$$R = |R'| = 1 - T = \frac{(k_1 - k_2)^2}{(k_1 + k_2)^2}$$

In the limit of large energies $E \gg V_0$, we have $k_1 \approx k_2$ and the classical result $T = 1$, $R = 0$ is recovered.

Thus there is a finite probability for a particle with an energy larger than the step height to be reflected.

Classical Limit

The result obtained for R depends only on the ratio E/V_0. This seems superficially to violate the correspondence principle, since we obtain a finite probability of reflection regardless of the value of Planck's constant or the mass of the particle. For example, we seem to predict that when a marble rolls to the edge of a table, there can be a large probability that it is reflected back rather than falling off. Consistency with classical mechanics is restored by eliminating the unphysical assumption that the step potential is discontinuous. When the step function is replaced with a ramp that spans some finite distance w, the probability of reflection approaches zero in the limit $wk \to \inf$, where k is the wavenumber of the particle.

Relativistic Calculation

The relativistic calculation of a free particle colliding with a step potential can be obtained using relativistic quantum mechanics. For the case of 1/2 fermions, like electrons and neutrinos, the solutions of the Dirac equation for high energy barriers produce transmission and reflection coefficients that are not bounded. This phenomenon is known as the Klein paradox. The apparent paradox disappears in the context of quantum field theory.

Applications

The Heaviside step potential mainly serves as an exercise in introductory quantum mechanics, as the solution requires understanding of a variety of quantum mechanical concepts: wavefunction normalization, continuity, incident/reflection/transmission amplitudes, and probabilities.

A similar problem to the one considered appears in the physics of normal-metal superconductor interfaces. Quasiparticles are scattered at the pair potential which in the simplest model may be assumed to have a step-like shape. The solution of the Bogoliubov-de Gennes equation resembles that of the discussed Heaviside-step potential. In the superconductor normal-metal case this gives rise to Andreev reflection.

ONE-DIMENSIONAL LINEAR HARMONIC OSCILLATOR

The harmonic oscillator is an extremely important physics problem. Many potentials look like a harmonic oscillator near their minimum. This is the first non-constant potential for which we will solve the Schrödinger Equation.

The harmonic oscillator Hamiltonian is given by:

$$H = \frac{p^2}{2m} + \frac{1}{2}kx^2$$

which makes the Schrödinger Equation for energy eigenstates:

$$\frac{-\hbar^2}{2m}\frac{d^2u}{dx^2} + \frac{1}{2}kx^2u = Eu.$$

Note that this potential also has a Parity symmetry. The potential is unphysical because it does not go to zero at infinity, however, it is often a very good approximation, and this potential can be solved exactly.

It is standard to remove the spring constant k from the Hamiltonian, replacing it with the classical oscillator frequency.

$$\omega = \sqrt{\frac{k}{m}}$$

The Harmonic Oscillator Hamiltonian becomes:

$$H = \frac{p^2}{2m} + \frac{1}{2}m\omega^2x^2$$

The differential equation to be solved is:

$$\frac{-\hbar^2}{2m}\frac{d^2u}{dx^2}+\frac{1}{2}m\omega^2x^2u=Eu.$$

To solve the Harmonic Oscillator equation, we will first change to dimensionless variables, then find the form of the solution for $x \to \pm\infty$, then multiply that solution by a polynomial, derive a recursion relation between the coefficients of the polynomial, show that the polynomial series must terminate if the solutions are to be normalizable, derive the energy eigenvalues, then finally derive the functions that are solutions.

The energy eigenvalues are:

$$E_n=\left(n+\frac{1}{2}\right)\hbar\omega$$

For $n=0,1,2,\ldots$ There are a countably infinite number of solutions with equal energy spacing. We have been forced to have quantized energies by the requirement that the wave functions be normalizable.

The ground state wave function is:

$$u_0(x)=\left(\frac{m\omega}{\pi\hbar}\right)^{\frac{1}{4}}e^{-m\omega x^2/2\hbar}$$

This is a Gaussian (minimum uncertainty) distribution. Since the HO potential has a parity symmetry, the solutions either have even or odd parity. The ground state is even parity.

The first excited state is an odd parity state, with a first order polynomial multiplying the same Gaussian:

$$u_1(x)=\left(\frac{m\omega}{\pi\hbar}\right)^{\frac{1}{4}}\sqrt{\frac{2m\omega}{\hbar}}xe^{-m\omega x^2/2\hbar}$$

The second excited state is even parity, with a second order polynomial multiplying the same Gaussian:

$$u_2(x)=C\left(1-2\frac{m\omega x^2}{\hbar}\right)e^{-m\omega x^2/2\hbar}$$

Note that n is equal to the number of zeros of the wavefunction. This is a common trend. With more zeros, a wavefunction has more curvature and hence more kinetic energy.

The general solution can be written as:

$$u_n(x) = \sum_{k=0}^{\infty} a_k y^k e^{-y^2/2}$$

with the coefficients determined by the recursion relation:

$$a_{k+2} = \frac{2(k-n)}{(k+1)(k+2)} a_k$$

and the dimensionless variable y given by.

$$y = \sqrt{\frac{m\omega}{\hbar}} x$$

The series terminates with the last nonzero term having $k = n$.

References

- Todd Wimpfheimer, A Particle in a Box Laboratory Experiment Using Everyday Compounds, Journal of Laboratory Chemical Education, Vol. 3 No. 2, 2015, pp. 19-21. Doi:10.5923/j.jlce.20150302.01
- Bound+State: thefreedictionary.com, Retrieved 9 June, 2020
- *Zandonella, Catherine.* "Rice-sized laser, powered one electron at a time, bodes well for quantum computing". *Princeton University. Princeton University.* Retrieved 8 November 2016
- *Miessler, G. L. (2013). Inorganic chemistry (5 ed.). Boston: Pearson. pp. 235–236.* ISBN 978-0321811059
- D. Branson. 'The correspondence principle and scattering from potential steps', American Journal of Physics, Vol.47, 1101-1102, 1979. http://www.deepdyve.com/lp/american-association-of-physics-teachers/correspondence-principle-and-scattering-from-potential-steps-tKM85ATfDZ/1

4

Three-Dimensional Potentials

Three-Dimensional potentials include three-dimensional box, spherically symmetric potentials, three-dimensional harmonic oscillator, etc. The particle in a box model deals with the movement of free particles in space surrounded by impenetrable barriers. This chapter discusses all the concepts related to three-dimensional potentials in detail.

THREE-DIMENSIONAL BOX

The quantum particle in the 1D box problem can be expanded to consider a particle within a higher dimensions as demonstrated elsewhere for a quantum particle in a 2D box. Here we continue the expansion into a particle trapped in a 3D box with three lengths L_x, L_y, and L_z. As with the other systems, there is no force (i.e., no potential) acting on the particles *inside* the box.

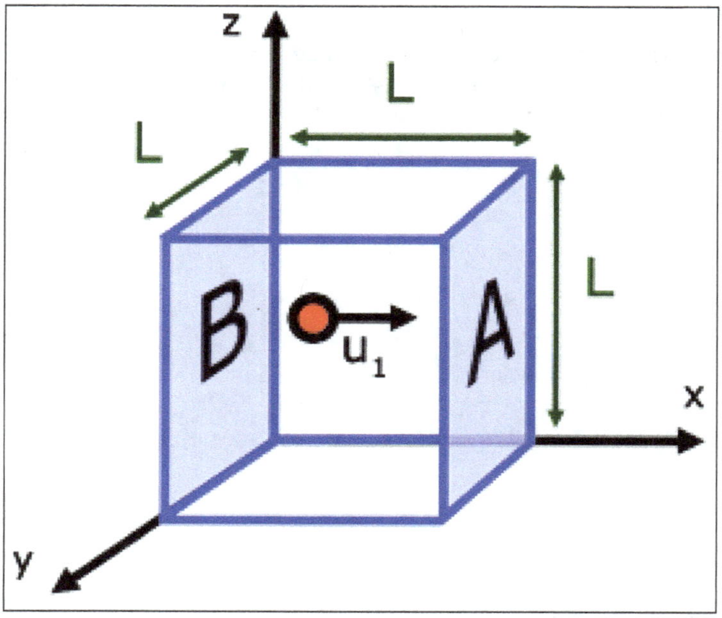

A particle in a 3-D box scheme with equal lengths on all three dimensions.

The potential for the particle inside the box:

$$V(\vec{r}\,) = 0$$

- $0 \le x \le L_x$.
- $0 \le y \le L_y$.
- $0 \le z \le L_z$.
- $L_x < x < 0$.
- $L_y < y < 0$.
- $L_z < z < 0$.

$\vec{r}$ is the vector with all three components along the three axes of the 3-D box: $\vec{r} = L_x\hat{x} + L_y\hat{y} + L_z\,\hat{z}$. When the potential energy is infinite, then the wavefunction equals zero. When the potential energy is zero, then the wavefunction obeys the Time-Independent Schrodinger Equation.

$$-\frac{\hbar^2}{2m}\nabla^2\psi(r) + V(r)\psi(r) = E\psi(r)$$

Since we are dealing with a 3-dimensional figure, we need to add the 3 different axes into the Schrondinger equation:

$$-\frac{\hbar^2}{2m}\left(\frac{d^2\psi(r)}{dx^2} + \frac{d^2\psi(r)}{dy^2} + \frac{d^2\psi(r)}{dz^2}\right) = E\psi(r)$$

The easiest way in solving this partial differential equation is by having the wavefunction equal to a product of individual function for each independent variable (e.g., the Separation of Variables technique):

$$\psi(x, y, z) = X(x)Y(y)Z(z)$$

Now each function has its own variable:

- X(x) is a function for variable x only.
- Y(y) function of variable y only.
- Z(z) function of variable z only.

Now substitute equation $\psi(x, y, z) = X(x)Y(y)Z(z)$ into equation:

$-\frac{\hbar^2}{2m}\left(\frac{d^2\psi(r)}{dx^2} + \frac{d^2\psi(r)}{dy^2} + \frac{d^2\psi(r)}{dz^2}\right) = E\psi(r)$ and divide it by the product: xyz:

$$\frac{d^2\psi}{dx^2} = YZ\frac{d^2X}{dx^2} \Rightarrow \frac{1}{X}\frac{d^2X}{dx^2}$$

$$\frac{d^2\psi}{dy^2} = XZ\frac{d^2Y}{dy^2} \Rightarrow \frac{1}{Y}\frac{d^2Y}{dy^2}$$

$$\frac{d^2\psi}{dz^2} = XY\frac{d^2Z}{dz^2} \Rightarrow \frac{1}{Z}\frac{d^2Z}{dz^2}$$

$$\left(-\frac{\hbar^2}{2mX}\frac{d^2X}{dx^2}\right)+\left(-\frac{\hbar^2}{2mY}\frac{d^2Y}{dy^2}\right)+\left(-\frac{\hbar^2}{2mZ}\frac{d^2Z}{dz^2}\right)=E$$

E is an energy constant, and is the sum of x, y, and z. For this to work, each term must equal its own constant. For example,

$$\frac{d^2X}{dx^2}+\frac{2m}{\hbar^2}\varepsilon_x X=0$$

Now separate each term in equation $\left(-\frac{\hbar^2}{2mX}\frac{d^2X}{dx^2}\right)+\left(-\frac{\hbar^2}{2mY}\frac{d^2Y}{dy^2}\right)+\left(-\frac{\hbar^2}{2mZ}\frac{d^2Z}{dz^2}\right)=E$ to equal zero:

$$\frac{d^2X}{dx^2}+\frac{2m}{\hbar^2}\varepsilon_x X=0$$

$$\frac{d^2Y}{dy^2}+\frac{2m}{\hbar^2}\varepsilon_y Y=0$$

$$\frac{d^2Z}{dz^2}+\frac{2m}{\hbar^2}\varepsilon_z Z=0$$

Now we can add all the energies together to get the total energy:

$$\varepsilon_x+\varepsilon_y+\varepsilon_z=E\,.$$

We have now reduced the 3D box into three particle in a 1D box problems:

$$\frac{d^2X}{dx^2}+\frac{2m}{\hbar 2}E_x X=0\approx\frac{d^2\psi}{dx^2}=-\frac{4\pi^2}{\lambda^2}\psi$$

Now the equations are very similar to a 1-D box and the boundary conditions are identical, i.e.,

$$n=1,2,..\infty$$

Use the normalization wavefunction equation for each variable:

$$\psi(\mathrm{x}) = \begin{cases} \sqrt{\dfrac{2}{L^2}} \sin \dfrac{n\pi x}{Lx} & \text{if } 0 \le \mathrm{x} \le \mathrm{L} \\ 0 & \text{if } L < \mathrm{x} < 0 \end{cases}$$

Normalization wavefunction equation for each variable:

$$X(x) = \sqrt{\frac{2}{L_x}}\left(\frac{n_x \pi x}{L_x}\right)$$

$$\mathrm{Y(y)} = \sqrt{\frac{2}{L_y}} \sin\left(\frac{\mathrm{n_y}\pi \mathrm{y}}{\mathrm{L_y}}\right)$$

$$Z(z) = \sqrt{\frac{2}{L_z}} \sin\left(\frac{n_z \pi z}{L_z}\right)$$

The limits of the three quantum numbers:

$$n_x = 1, 2, 3, \ldots \infty$$
$$n_y = 1, 2, 3, \ldots \infty$$
$$n_z = 1, 2, 3, \ldots \infty$$

For each constant use the de Broglie Energy equation:

$$\varepsilon_x = \frac{n_x^2 h^2}{8mL_x^2}$$

with $n_x = 1 \ldots \infty$:

Do the same for variables $\mathrm{n_y}$ and $\mathrm{n_z}$. Combine Equation $\psi(x, y, z) = X(x)Y(y)Z(z)$ with equations $X(x) = \sqrt{\frac{2}{L_x}}\left(\frac{n_x \pi x}{L_x}\right)$, $\mathrm{Y(y)} = \sqrt{\frac{2}{L_y}} \sin\left(\frac{\mathrm{n_y}\pi \mathrm{y}}{\mathrm{L_y}}\right)$, $Z(z) = \sqrt{\frac{2}{L_z}} \sin\left(\frac{n_z \pi z}{L_z}\right)$ to find the wavefunctions inside a 3D box.

$$\psi(r) = \sqrt{\frac{8}{V}} \sin\left(\frac{n_x \pi x}{L_x}\right) \sin\left(\frac{n_y \pi y}{L_y}\right) \sin\left(\frac{n_z \pi z}{L_z}\right)$$

with

$$V = \underbrace{L_x \times L_y \times L_z}_{volume\ of\ Box}$$

To find the Total Energy, add equation $\varepsilon_x = \dfrac{n_x^2 h^2}{8mL_x^2}$ and equation $\varepsilon_x + \varepsilon_y + \varepsilon_z = E$.

$$E_{n_x}, n_y, n_z = \frac{h^2}{8m}\left(\frac{n_x^2}{L_x^2} + \frac{n_y^2}{L_y^2} + \frac{n_z^2}{L_z^2}\right)$$

Notice the similarity between the energies a particle in a 3D box in the above equation, and a 1D box.

Degeneracy in a 3D Cube

The energy of the particle in a 3-D cube (i.e., a = L, b = L, and c = L) in the *ground state* is given by equation $E_{n_x}, n_y, n_z = \dfrac{h^2}{8m}\left(\dfrac{n_x^2}{L_x^2} + \dfrac{n_y^2}{L_y^2} + \dfrac{n_z^2}{L_z^2}\right)$ with $n_x = 1$, $n_y = 1$, and $n_z = 1$. This energy ($E_{1,1,1}$) is hence:

$$E_{1,1,1} = \frac{3h^2}{8mL^2}$$

The ground state has only one wavefunction and no other state has this specific energy; the ground state and the energy level are said to be non-degenerate. However, in the 3-D cubical box potential the energy of a state depends upon the sum of the squares of the quantum numbers equation $\psi(r) = \sqrt{\dfrac{8}{V}}\sin\left(\dfrac{n_x \pi x}{L_x}\right)\sin\left(\dfrac{n_y \pi y}{L_y}\right)\sin\left(\dfrac{n_z \pi z}{L_z}\right)$. The particle having a particular value of energy in the excited state MAY has several different stationary states or wavefunctions. If so, these states and energy eigenvalues are said to be degenerate.

For the first excited state, three combinations of the quantum numbers (n_x, n_y, n_z) are (2,1,1), (1,2,1), (1,1,2). The sum of squares of the quantum numbers in each combination is same (equal to 6). Each wavefunction has same energy:

$$E_{2,1,1} = E_{1,2,1} = E_{1,1,2} = \frac{6h^2}{8mL^2}$$

Corresponding to these combinations three different wavefunctions and three different states are possible. Hence, the first excited state is said to be three-fold or triply degenerate. The number of independent wavefunctions for the stationary states of an energy level is called as the degree of degeneracy of the energy level. The value of energy levels with the corresponding combinations and sum of squares of the quantum numbers:

$$n^2 = n_x^2 + n_y^2 + n_z^2$$

as well as the degree of degeneracy are depicted in table.

Table: Degeneracy properties of the particle in a 3-D cube with a = L, b = L, and c = L.

$n_x^2 + n_y^2 + n_x^2$	Combinations of Degeneracy (n_x, n_y, n_z)	Total Energy (E_{nx}, n_y, n_z)	Degree of Degeneracy
3	(1,1,1)	$\frac{3h^2}{8mL^2}$	1
6	(2,1,1) (1,2,1) (1,1,2)	$\frac{6h^2}{8mL^2}$	3
9	(2,2,1) (1,2,2) (2,1,2)	$\frac{9h^2}{8mL^2}$	3
11	(3,1,1) (1,3,1) (1,1,3)	$\frac{11h^2}{8mL^2}$	3
12	(2,2,2)	$\frac{12h^2}{8mL^2}$	1
14	(3,2,1) (3,1,2) (2,3,1) (2,1,3) (1,3,2) (1,2,3)	$\frac{14h^2}{8mL^2}$	6
17	(2,2,3) (3,2,2) (2,3,2)	$\frac{17h^2}{8mL^2}$	3
18	(1,1,4) (1,4,1) (4,1,1)	$\frac{18h^2}{8mL^2}$	3
19	(1,3,3) (3,1,3) (3,3,1)	$\frac{19h^2}{8mL^2}$	3
21	(1,2,4) (1,4,2) (2,1,4) (2,4,1) (4,1,2) (4,2,1)	$\frac{21h^2}{8mL^2}$	6

SPHERICALLY SYMMETRIC POTENTIALS

An important problem in quantum mechanics is that of a particle in a spherically symmetric potential, i.e., a potential that depends only on the distance between the particle and a defined center point. In particular, if the particle in question is an electron and the potential is derived from Coulomb's law, then the problem can be used to describe a hydrogen-like (one-electron) atom (or ion).

In the general case, the dynamics of a particle in a spherically symmetric potential are governed by a Hamiltonian of the following form:

$$\hat{H} = \frac{\hat{p}^2}{2m_0} + V(r)$$

where m_0 is the mass of the particle, $\hat{p}$ is the momentum operator, and the potential $V(r)$ depends only on r, the modulus of the radius vector **r**. The quantum mechanical wavefunctions and energies (eigenvalues) are found by solving the Schrödinger equation with this Hamiltonian. Due to the spherical symmetry of the system, it is natural to use spherical coordinates r, θ and ϕ. When this is done, the time-independent Schrödinger equation for the system is separable, allowing the angular problems to be dealt with easily, and leaving an ordinary differential equation in r to determine the energies for the particular potential $V(r)$ under discussion.

Structure of the Eigenfunctions

The eigenstates of the system have the form:

$$\psi(r,\theta,\phi) = R(r)\Theta(\theta)\Phi(\phi)$$

in which the spherical polar angles θ and φ represent the colatitude and azimuthal angle, respectively. The last two factors of ψ are often grouped together as spherical harmonics, so that the eigenfunctions take the form:

$$\psi(r,\theta,\phi) = R(r)Y_{lm}(\theta,\phi)$$

The differential equation which characterizes the function R(r) is called the radial equation.

Derivation of the Radial Equation

The kinetic energy operator in spherical polar coordinates is:

$$\frac{\hat{p}^2}{2m_0} = -\frac{\hbar^2}{2m_0}\nabla^2 = -\frac{\hbar^2}{2m_0 r^2}\left[\frac{\partial}{\partial r}\left(r^2\frac{\partial}{\partial r}\right) - \hat{l}^2\right].$$

The spherical harmonics satisfy:

$$\hat{l}^2 Y_{lm}(\theta,\phi) \equiv \left\{-\frac{1}{\sin^2\theta}\left[\sin\theta\frac{\partial}{\partial\theta}\left(\sin\theta\frac{\partial}{\partial\theta}\right) + \frac{\partial^2}{\partial\phi^2}\right]\right\} Y_{lm}(\theta,\phi) = l(l+1)Y_{lm}(\theta,\phi)$$

Substituting this into the Schrödinger equation we get a one-dimensional eigenvalue equation,

$$\left\{-\frac{\hbar^2}{2m_0 r^2}\frac{d}{dr}\left(r^2\frac{d}{dr}\right) + \frac{\hbar^2 l(l+1)}{2m_0 r^2} + V(r)\right\} R(r) = ER(r).$$

Relationship with 1-D Schrödinger Equation

Note that the first term in the kinetic energy can be rewritten:

$$T_r \equiv \frac{1}{r^2}\frac{d}{dr}r^2\frac{d}{dr} = \frac{1}{r}\frac{d^2}{dr^2}r.$$

If subsequently the substitution $u(r) \overset{\text{def}}{=} rR(r)$ is made into:

$$-\frac{\hbar^2}{2m_0 r}\frac{d^2}{dr^2}\left(rR(r)\right)+\frac{\hbar^2 l(l+1)}{2m_0 r^2}R(r)+V(r)R(r)=ER(r),$$

the radial equation becomes:

$$-\frac{\hbar^2}{2m_0}\frac{d^2 u}{dr^2}+V_{\text{eff}}(r)u(r)=Eu(r)$$

which is precisely a Schrödinger equation for the function u(r) with an effective potential given by:

$$V_{\text{eff}}(r)=V(r)+\frac{\hbar^2 l(l+1)}{2m_0 r^2},$$

where the radial coordinate r ranges from 0 to ∞. The correction to the potential V(r) is called the centrifugal barrier term.

Solutions for Potentials of Interest

Five special cases arise, of special importance:

- $V(r) = 0$, or solving the vacuum in the basis of spherical harmonics, which serves as the basis for other cases.
- $V(r)=V_0$ (finite) for $r<r_0$ and 0 elsewhere, or a particle in the spherical equivalent of the square well, useful to describe scattering and bound states in a nucleus or quantum dot.
- As the previous case, but with an infinitely high jump in the potential on the surface of the sphere.
- $V(r) \sim r^2$ for the three-dimensional isotropic harmonic oscillator.
- $V(r) \sim 1/r$ to describe bound states of hydrogen-like atoms.

We outline the solutions in these cases, which should be compared to their counterparts in cartesian coordinates, cf. particle in a box. This topic relies heavily on Bessel functions and Laguerre polynomials.

Vacuum Case

Let us now consider $V(r) = 0$ (if V_0, replace everywhere E with $E-V_0$). Introducing the dimensionless variable:

$$\rho \overset{\text{def}}{=} kr, \qquad k \overset{\text{def}}{=} \sqrt{\frac{2m_0 E}{\hbar^2}}$$

the equation becomes a Bessel equation for J defined by $J(\rho) \stackrel{\text{def}}{=} \sqrt{\rho}R(r)$ (whence the notational choice of J):

$$\rho^2 \frac{d^2J}{d\rho^2} + \rho\frac{dJ}{d\rho} + \left[\rho^2 - \left(l + \frac{1}{2}\right)^2\right]J = 0$$

which regular solutions for positive energies are given by so-called Bessel functions of the first kind':

$J_{l+1/2}(\rho)$ so that the solutions written for R are the so-called Spherical Bessel function:

$$R(r) = j_l(kr) \stackrel{\text{def}}{=} \sqrt{\pi / (2kr)} J_{l+1/2}(kr).$$

The solutions of Schrödinger equation in polar coordinates for a particle of mass m_0 in vacuum are labelled by three quantum numbers: discrete indices l and m, and k varying continuously in [0, ∞):

$$\psi(\mathbf{r}) = j_l(kr)Y_{lm}(\theta,\phi):$$

where $k \stackrel{\text{def}}{=} \sqrt{2m_0E}/\hbar$, j_l are the spherical Bessel functions and Y_{lm} are the spherical harmonics.

These solutions represent states of definite angular momentum, rather than of definite (linear) momentum, which are provided by plane waves $\exp(i\mathbf{k}\cdot\mathbf{r})$.

Sphere with Finite "Square" Potential

Let us now consider the potential $V(r) = V_0$ for $r < r_0$ and $V(r) = 0$ elsewhere. That is, inside a sphere of radius r_0 the potential is equal to V_0 and it is zero outside the sphere. A potential with such a finite discontinuity is called a *square potential.*

We first consider bound states, i.e., states which display the particle mostly inside the box (confined states). Those have an energy *E* less than the potential outside the sphere, i.e., they have negative energy, and we shall see that there are a discrete number of such states, which we shall compare to positive energy with a continuous spectrum, describing scattering on the sphere (of unbound states). Also worth noticing is that unlike Coulomb potential, featuring an infinite number of discrete bound states, the spherical square well has only a finite (if any) number because of its finite range (if it has finite depth).

The resolution essentially follows that of the vacuum with normalization of the total wavefunction added, solving two Schrödinger equations — inside and outside the sphere — of the previous kind, i.e., with constant potential. Also the following constraints hold:

- The wavefunction must be regular at the origin.
- The wavefunction and its derivative must be continuous at the potential discontinuity.

- The wavefunction must converge at infinity.

The first constraint comes from the fact that Neumann *N* and Hankel *H* functions are singular at the origin. The physical argument that ψ must be defined everywhere selected Bessel function of the first kind *J* over the other possibilities in the vacuum case. For the same reason, the solution will be of this kind inside the sphere:

$$R(r) = Aj_l\left(\sqrt{\frac{2m_0(E-V_0)}{\hbar^2}}r\right), \qquad r < r_0$$

with A a constant to be determined later. Note that for bound states, $V_0 < E < 0$.

Bound states bring the novelty as compared to the vacuum case that E is now negative (in the vacuum it was to be positive). This, along with third constraint, selects Hankel function of the first kind as the only converging solution at infinity (the singularity at the origin of these functions does not matter since we are now outside the sphere):

$$R(r) = Bh_l^{(1)}\left(i\sqrt{\frac{-2m_0E}{\hbar^2}}r\right), \qquad r > r_0$$

Second constraint on continuity of ψ at $r = r_0$ along with normalization allows the determination of constants *A* and *B*. Continuity of the derivative (or logarithmic derivative for convenience) requires quantization of energy.

Sphere with Infinite "Square" Potential

In case where the potential well is infinitely deep, so that we can take $V_0 = 0$ inside the sphere and ∞ outside, the problem becomes that of matching the wavefunction inside the sphere (the spherical Bessel functions) with identically zero wavefunction outside the sphere. Allowed energies are those for which the radial wavefunction vanishes at the boundary. Thus, we use the zeros of the spherical Bessel functions to find the energy spectrum and wavefunctions. Calling $u_{l,k}$ the kth zero of j_l, we have:

$$E_{kl} = \frac{u_{l,k}^2\hbar^2}{2m_0r_0^2}$$

So that one is reduced to the computations of these zeros $u_{l,k}$ typically by using a table or calculator, as these zeros are not solvable for the general case.

In the special case $l = 0$ (spherical symmetric orbitals), the spherical Bessel function is $j_0(x) = \dfrac{\sin x}{x}$ which zeros can be easily given as $u_{0,k} = k\pi$. Their energy eigenvalues are thus:

$$E_{k0} = \frac{(k\pi)^2\hbar^2}{2m_0r_0^2} = \frac{k^2h^2}{8m_0r_0^2}$$

3D Isotropic Harmonic Oscillator

The potential of a 3D isotropic harmonic oscillator is:

$$V(r) = \frac{1}{2} m_0 \omega^2 r^2.$$

In this topic it is shown that an N-dimensional isotropic harmonic oscillator has the energies:

$$E_n = \hbar\omega\left(n + \frac{N}{2}\right) \quad \text{with} \quad n = 0, 1, \ldots, \infty,$$

i.e., n is a non-negative integral number; ω is the (same) fundamental frequency of the N modes of the oscillator. In this case N = 3, so that the radial Schrödinger equation becomes,

$$\left[-\frac{\hbar^2}{2m_0}\frac{d^2}{dr^2} + \frac{\hbar^2 l(l+1)}{2m_0 r^2} + \frac{1}{2} m_0 \omega^2 r^2 - \hbar\omega\left(n + \tfrac{3}{2}\right) \right] u(r) = 0.$$

Introducing:

$$\gamma \equiv \frac{m_0 \omega}{\hbar}$$

and recalling that $u(r) = rR(r)$, we will show that the radial Schrödinger equation has the normalized solution,

$$R_{n,l}(r) = N_{nl}\, r^l\, e^{-\frac{1}{2}\gamma r^2}\, L_{\frac{1}{2}(n-l)}^{(l+\frac{1}{2})}(\gamma r^2),$$

where the function $L_k^{(\alpha)}(\gamma r^2)$ is a generalized Laguerre polynomial in γr^2 of order k (i.e., the highest power of the polynomial is proportional to $\gamma^k r^{2k}$).

The normalization constant N_{nl} is,

$$N_{nl} = \left[\frac{2^{n+l+2}\, \gamma^{l+\frac{3}{2}}}{\pi^{\frac{1}{2}}} \right]^{\frac{1}{2}} \left[\frac{[\frac{1}{2}(n-l)]!\,[\frac{1}{2}(n+l)]!}{(n+l+1)!} \right]^{\frac{1}{2}}.$$

The eigenfunction $R_{n,l}(r)$ belongs to energy E_n and is to be multiplied by the spherical harmonic $Y_{lm}(\theta,\phi)$, where:

$$l = n, n-2, \ldots, l_{\min} \quad \text{with} \quad l_{\min} = \begin{cases} 1 & \text{if } n \text{ odd} \\ 0 & \text{if } n \text{ even} \end{cases}$$

Hydrogen-like Atoms

A hydrogenic (hydrogen-like) atom is a two-particle system consisting of a nucleus and an electron. The two particles interact through the potential given by Coulomb's law:

$$V(r) = -\frac{1}{4\pi\epsilon_0}\frac{Ze^2}{r}$$

where:

- ε_o is the permittivity of the vacuum.
- Z is the atomic number (eZ is the charge of the nucleus).
- e is the elementary charge (charge of the electron).
- r is the distance between the electron and the nucleus.

The mass m_o, introduced above, is the reduced mass of the system. Because the electron mass is about 1836 times smaller than the mass of the lightest nucleus (the proton), the value of m_o is very close to the mass of the electron m_e for all hydrogenic atoms. In the remaining of the topic we make the approximation $m_o = m_e$. Since m_e will appear explicitly in the formulas it will be easy to correct for this approximation if necessary.

In order to simplify the Schrödinger equation, we introduce the following constants that define the atomic unit of energy and length, respectively,

$$E_h = m_e\left(\frac{e^2}{4\pi\varepsilon_0\hbar}\right)^2 \quad \text{and} \quad a_0 = \frac{4\pi\varepsilon_0\hbar^2}{m_e e^2}.$$

Substitute $y = Zr/a_0$ and $W = E/(Z^2E_h)$ into the radial Schrödinger equation given above. This gives an equation in which all natural constants are hidden,

$$\left[-\frac{1}{2}\frac{d^2}{dy^2}+\frac{1}{2}\frac{l(l+1)}{y^2}-\frac{1}{y}\right]u_l = Wu_l.$$

Two classes of solutions of this equation exist: (i) W is negative, the corresponding eigenfunctions are square integrable and the values of W are quantized (discrete spectrum). (ii) W is non-negative. Every real non-negative value of W is physically allowed (continuous spectrum), the corresponding eigenfunctions are non-square integrable. In the remaining part of this topic only class (i) solutions will be considered. The wavefunctions are known as bound states, in contrast to the class (ii) solutions that are known as scattering states.

For negative W the quantity $\alpha \equiv 2\sqrt{-2W}$ is real and positive. The scaling of y, i.e., substitution of $x \equiv \alpha y$ gives the Schrödinger equation:

$$\left[\frac{d^2}{dx^2}-\frac{l(l+1)}{x^2}+\frac{2}{\alpha x}-\frac{1}{4}\right]u_l = 0, \quad \text{with } x \geq 0.$$

For $x \to \infty$ the inverse powers of x are negligible and a solution for large x is $\exp[-x/2]$ The other solution, $\exp[x/2]$ is physically non-acceptable. For $x \to 0$ the inverse square power dominates and a solution for small x is x^{l+1}. The other solution, x^{-l}, is physically non-acceptable. Hence, to obtain a full range solution we substitute:

$$u_l(x) = x^{l+1} e^{-x/2} f_l(x).$$

The equation for $f_l(x)$ becomes,

$$\left[x\frac{d^2}{dx^2} + (2l+2-x)\frac{d}{dx} + (\nu - l - 1) \right] f_l(x) = 0 \quad \text{with} \quad \nu = (-2W)^{-\frac{1}{2}}.$$

Provided $\nu - l - 1$ is a non-negative integer, say k, this equation has polynomial solutions written as:

$$L_k^{(2l+1)}(x), \qquad k = 0, 1, \ldots,$$

which are generalized Laguerre polynomials of order k. We will take the convention for generalized Laguerre polynomials of Abramowitz and Stegun. Note that the Laguerre polynomials given in many quantum mechanical textbooks, for instance the book of Messiah, are those of Abramowitz and Stegun multiplied by a factor $(2l + 1 + k)$.

The energy becomes:

$$W = -\frac{1}{2n^2} \quad \text{with} \quad n \equiv k + l + 1.$$

The principal quantum number n satisfies $n \geq l+1$, or $l \leq n-1$ Since $\alpha = 2/n$, the total radial wavefunction is:

$$R_{nl}(r) = N_{nl} \left(\frac{2Zr}{na_0} \right)^l e^{-\frac{Zr}{na_0}} L_{n-l-1}^{(2l+1)} \left(\frac{2Zr}{na_0} \right),$$

with normalization constant,

$$N_{nl} = \left[\left(\frac{2Z}{na_0} \right)^3 \cdot \frac{(n-l-1)!}{2n[(n+l)!]^3} \right]^{\frac{1}{2}}$$

which belongs to the energy,

$$E = -\frac{Z^2}{2n^2} E_{\mathrm{h}}, \qquad n = 1, 2, \ldots.$$

In the computation of the normalization constant use was made of the integral:

$$\int_0^\infty x^{2l+2} e^{-x} \left[L_{n-l-1}^{(2l+1)}(x) \right]^2 dx = \frac{2n(n+l)!}{(n-l-1)!}.$$

THREE-DIMENSIONAL HARMONIC OSCILLATOR

The 3D harmonic oscillator can also be separated in Cartesian coordinates. For the case of a central potential, $V = \frac{1}{2} m\omega^2 r^2$, this problem can also be solved nicely in spherical coordinates using rotational symmetry. The cartesian solution is easier and better for counting states though.

Lets assume the central potential so we can compare to our later solution. We could have three different spring constants and the solution would be as simple. The Hamiltonian is:

$$H = \frac{p^2}{2m} + \frac{1}{2} m\omega^2 r^2$$

$$H = \frac{p_x^2}{2m} + \frac{1}{2} m\omega^2 x^2 + \frac{p_y^2}{2m} + \frac{1}{2} m\omega^2 y^2 + \frac{p_z^2}{2m} + \frac{1}{2} m\omega^2 z^2$$

$$H = H_x + H_y + H_z$$

The problem separates nicely, giving us three independent harmonic oscillators:

$$E = \left(n_x + n_y + n_z + \frac{3}{2} \right) \hbar\omega$$

$$\psi_{nx,ny,nz}(x,y,z) = u_{nx}(x) u_{ny}(y) u_{nz}(z)$$

This was really easy.

This problem has a different Fermi surface in n-space than did the particle in a box. The boundary between filled and unfilled energy levels is a plane defined by:

$$n_x + n_y + n_z = \frac{E_F}{\hbar\omega} - \frac{3}{2}$$

5

Angular Momentum

Angular momentum is the product of inertia and angular velocity and defines the quantity of rotation of a body. Spin is a form of angular momentum which is carried by elementary and composite particles, and atomic nuclei. This chapter closely examines the concepts of angular momentum to provide an easy understanding of the subject.

Angular momentum plays a central role in both classical and quantum mechanics. In classical mechanics, all isolated systems conserve angular momentum (as well as energy and linear momentum); this fact reduces considerably the amount of work required in calculating trajectories of planets, rotation of rigid bodies, and many more.

Similarly, in quantum mechanics, angular momentum plays a central role in understanding the structure of atoms, as well as other quantum problems that involve rotational symmetry.

Like other observable quantities, angular momentum is described in QM by an operator. This is in fact a vector operator, similar to momentum operator. However, as we will shortly see, contrary to the linear momentum operator, the three components of the angular momentum operator do not commute.

In QM, there are several angular momentum operators: the total angular momentum (usually denoted by $\vec{J}$), the orbital angular momentum (usually denoted by $\vec{L}$) and the intrinsic, or spin angular momentum (denoted by $\vec{S}$). This last one (spin) has no classical analogue. Confusingly, the term "angular momentum" can refer to either the total angular momentum, or to the orbital angular momentum.

The classical definition of the orbital angular momentum, $\vec{\mathrm{L}} = \vec{\mathrm{r}} \times \vec{\mathrm{p}}$ can be carried directly to QM by reinterpreting $\vec{\mathrm{r}}$ and $\vec{\mathrm{p}}$ as the operators associated with the position and the linear momentum.

The spin operator, **S**, represents another type of angular momentum, associated with "intrinsic rotation" of a particle around an axis; Spin is an intrinsic property of a particle (nearly all elementary particles have spin), that is unrelated to its spatial motion.

The existence of spin angular momentum is inferred from experiments, such as the Stern-Gerlach experiment, in which particles are observed to possess angular momentum that cannot be accounted for by orbital angular momentum alone.

The total angular momentum, **J**, combines both the spin and orbital angular momentum of a particle (or a system), namely $\vec{J} = \vec{L} + \vec{S}$.

ANGULAR MOMENTUM DIAGRAMS

In quantum mechanics and its applications to quantum many-particle systems, notably quantum chemistry, angular momentum diagrams, or more accurately from a mathematical viewpoint angular momentum graphs, are a diagrammatic method for representing angular momentum quantum states of a quantum system allowing calculations to be done symbolically. More specifically, the arrows encode angular momentum states in bra–ket notation and include the abstract nature of the state, such as tensor products and transformation rules.

The notation parallels the idea of Penrose graphical notation and Feynman diagrams. The diagrams consist of arrows and vertices with quantum numbers as labels, hence the alternative term "graphs". The sense of each arrow is related to Hermitian conjugation, which roughly corresponds to time reversal of the angular momentum states. The diagrammatic notation is a considerably large topic in its own right with a number of specialized features. They were developed primarily by Adolfas Jucys in the twentieth century.

Equivalence between Dirac Notation and Jucys Diagrams

Angular Momentum States

The quantum state vector of a single particle with total angular momentum quantum number j and total magnetic quantum number $m = j, j-1, \ldots, -j+1, -j$, is denoted as a ket $|j, m\rangle$. As a diagram this is a singleheaded arrow.

Symmetrically, the corresponding bra is $\langle j, m|$. In diagram form this is a doubleheaded arrow, pointing in the opposite direction to the ket.

In each case:

- The quantum numbers j, m are often labelled next to the arrows to refer to a specific angular momentum state.
- Arrowheads are almost always placed at the middle of the line, rather than at the tip.

- Equals signs "=" are placed between equivalent diagrams, exactly like for multiple algebraic expressions equal to each other.

The most basic diagrams are for kets and bras:

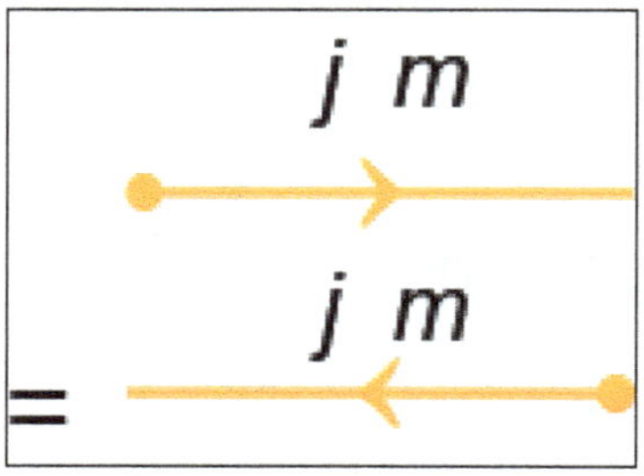

Ket $|j, m\rangle$.

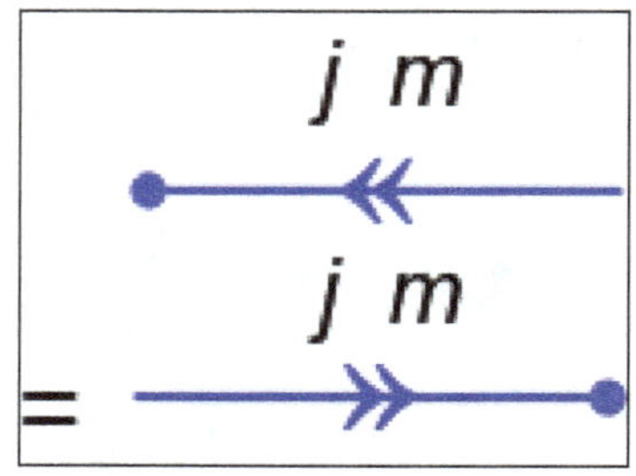

Bra $\langle j, m|$.

Arrows are directed to or from vertices, a state transforming according to:

- A standard representation is designated by an oriented line leaving a vertex.
- A contrastandard representation is depicted as a line entering a vertex.

As a general rule, the arrows follow each other in the same sense. In the contrastandard representation, the time reversal operator, denoted here by T, is used. It is unitary, which means the Hermitian conjugate $T^\dagger$ equals the inverse operator T^{-1}, that is $T^\dagger = T^{-1}$. It's action on the position operator leaves it invariant:

$$T\hat{\mathbf{x}}T^\dagger \quad \hat{\mathbf{x}}$$

but the linear momentum operator becomes negative,

$$T\hat{\mathbf{p}}T^\dagger = -\hat{\mathbf{p}}$$

and the spin operator becomes negative,

$$T\hat{\mathbf{S}}T^\dagger = -\hat{\mathbf{S}}$$

Since the orbital angular momentum operator is $\mathbf{L} = \mathbf{x} \times \mathbf{p}$, this must also become negative:

$$T\hat{\mathbf{L}}T^\dagger = -\hat{\mathbf{L}}$$

and therefore the total angular momentum operator $\mathbf{J} = \mathbf{L} + \mathbf{S}$ becomes negative:

$$T\hat{\mathbf{J}}T^\dagger = -\hat{\mathbf{J}}$$

Acting on an eigenstate of angular momentum $|j, m\rangle$, it can be shown that:

$$T|j,m\rangle \equiv |T(j,m)\rangle = (-1)^{j-m}|j,-m\rangle$$

The time-reversed diagrams for kets and bras are:

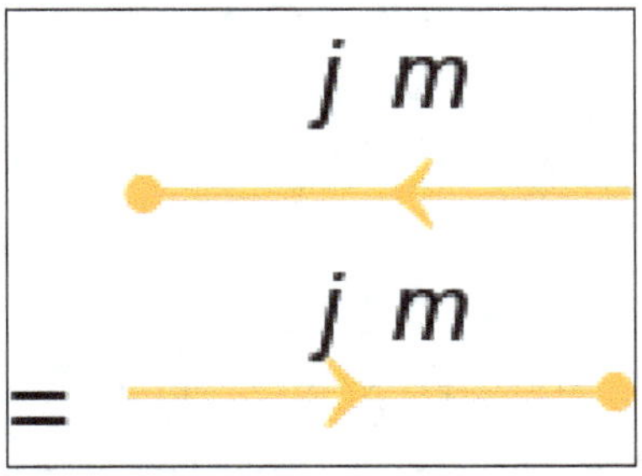

Time reversed ket $|j, m\rangle$.

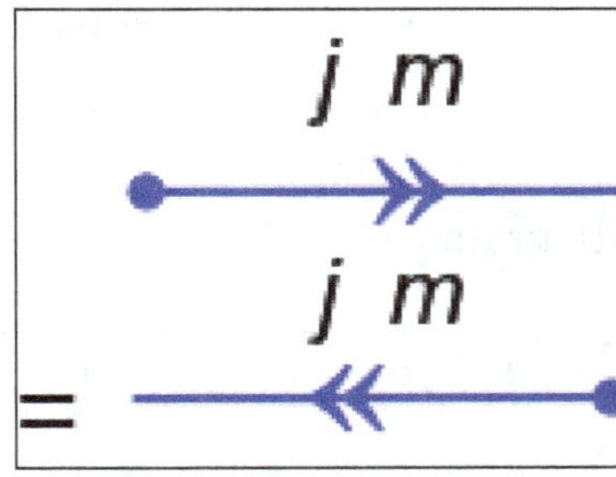

Time reversed bra $\langle j, m|$.

It is important to position the vertex correctly, as forward-time and reversed-time operators would become mixed up.

Inner Product

The inner product of two states $|j_1, m_1\rangle$ and $|j_2, m_2\rangle$ is:

$$\langle j_2, m_2 \mid j_1, m_1 \rangle = \delta_{j_1 j_2} \delta_{m_1 m_2}$$

and the diagrams are:

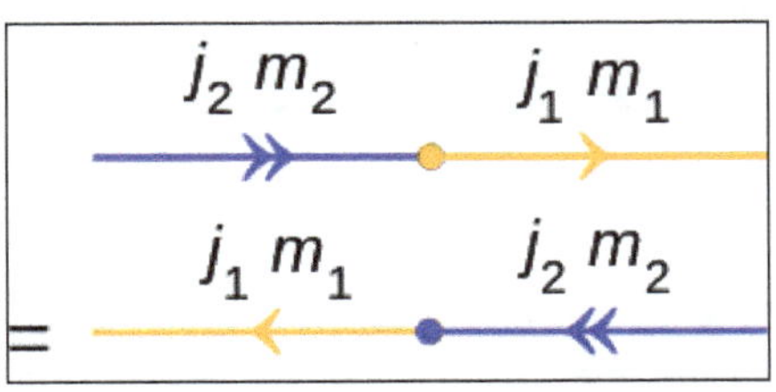

Inner product of $|j_1, m_1\rangle$ and $|j_2, m_2\rangle$, that is $\langle j_2, m_2|j_1, m_1\rangle$.

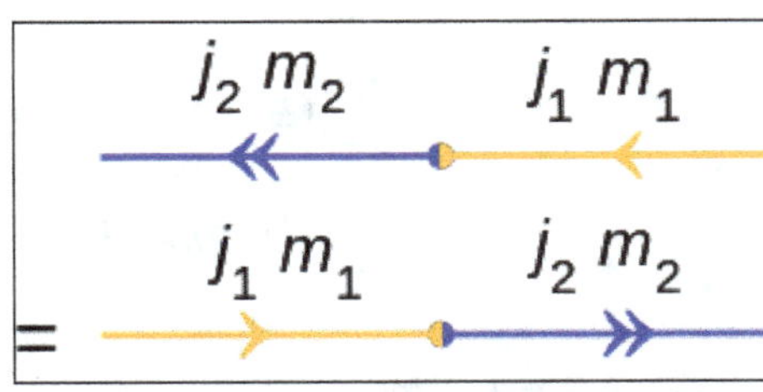

Time reversed equivalent.

For summations over the inner product, also known in this context as a contraction:

$$\sum_m \langle j, m \mid j, m \rangle = 2j + 1$$

it is conventional to denote the result as a closed circle labelled only by j, not m:

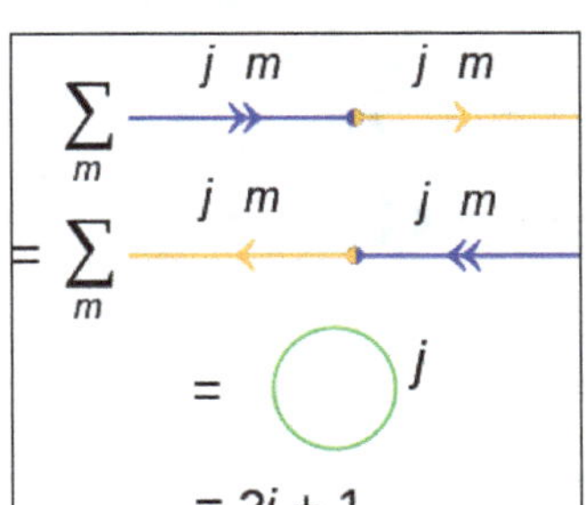

Inner product contraction.

Outer Products

The outer product of two states $|j_1, m_1\rangle$ and $|j_2, m_2\rangle$ is an operator:

$$|j_2,m_2\rangle\langle j_1,m_1|$$

and the diagrams are:

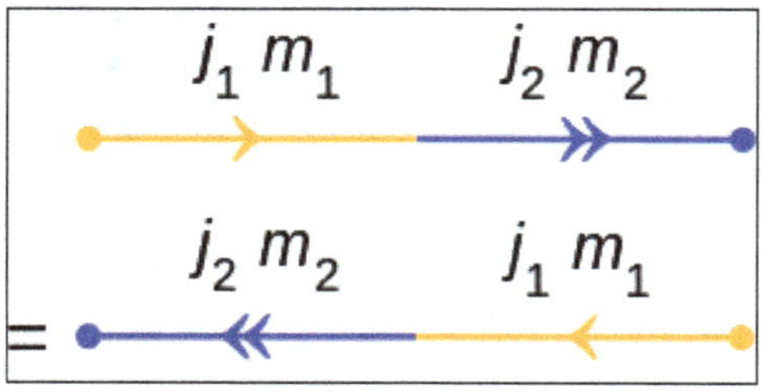

Outer product of $|j_1, m_1\rangle$ and $|j_2, m_2\rangle$, that is $|j_2, m_2\rangle\langle j_1, m_1|$.

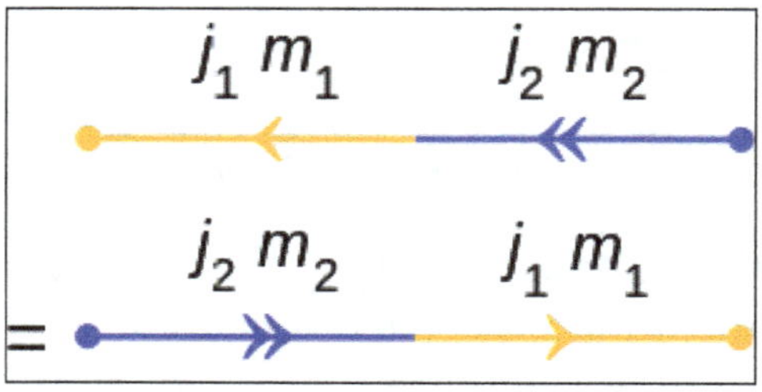

Time reversed equivalent.

For summations over the outer product, also known in this context as a contraction:

$$\begin{aligned}\sum_m |j,m\rangle\langle j,m| &= \sum_m |j,-m\rangle\langle j,-m| \\ &= \sum_m (-1)^{2(j-m)}\,|j,-m\rangle\langle j,-m| \\ &= \sum_m (-1)^{j-m}\,|j,-m\rangle\langle j,-m|(-1)^{j-m} \\ &= \sum_m T\,|j,m\rangle\langle j,m|\,T^\dagger\end{aligned}$$

where the result for $T|j, m\rangle$ was used, and the fact that m takes the set of values given above. There is no difference between the forward-time and reversed-time states for the outer product contraction, so here they share the same diagram, represented as one line without direction, again labelled by j only and not m:

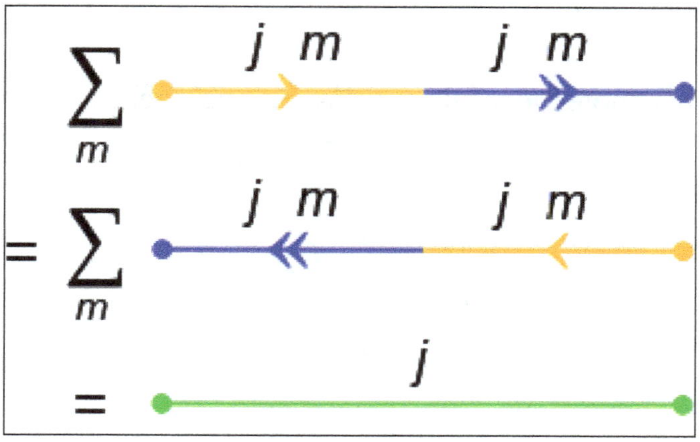

Outer product contraction.

Tensor Products

The tensor product ⊗ of n states $|j_1, m_1\rangle$, $|j_2, m_2\rangle$,... $|j_n, m_n\rangle$ is written:

$$\begin{aligned}|j_1,m_1,j_2,m_2,\ldots j_n,m_n\rangle &\equiv |j_1,m_1\rangle \otimes |j_2,m_2\rangle \otimes \cdots \otimes |j_n,m_n\rangle \\ &\equiv |j_1,m_1\rangle |j_2,m_2\rangle \cdots |j_n,m_n\rangle\end{aligned}$$

and in diagram form, each separate state leaves or enters a common vertex creating a "fan" of arrows - n lines attached to a single vertex.

Vertices in tensor products have signs (sometimes called "node signs"), to indicate the ordering of the tensor-multiplied states:

- A minus sign **(−)** indicates the ordering is *clockwise*, ↻.
- A plus sign **(+)** for *anticlockwise*, ↺.

Signs are of course not required for just one state, diagrammatically one arrow at a vertex. Sometimes curved arrows with the signs are included to show explicitly the sense of tensor multiplication, but usually just the sign is shown with the arrows left out.

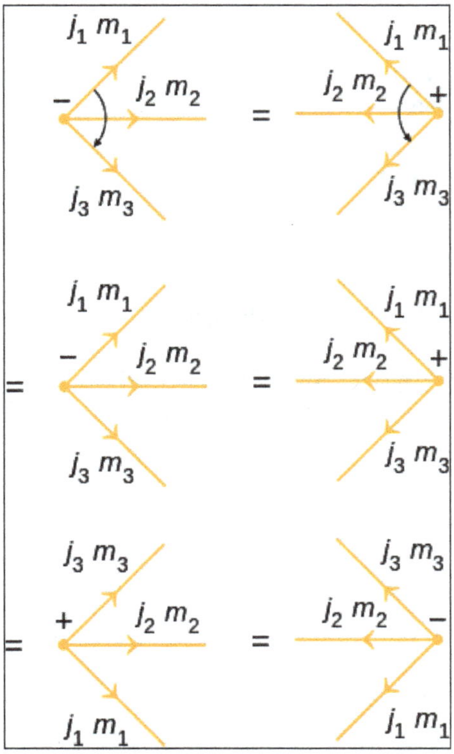

Tensor product of $|j_1, m_1\rangle$, $|j_2, m_2\rangle$, $|j_3, m_3\rangle$, that is $|j_1, m_1\rangle|j_2, m_2\rangle|j_3, m_3\rangle = |j_1, m_1, j_2, m_2, j_3, m_3\rangle$. Similarly for more than three angular momenta.

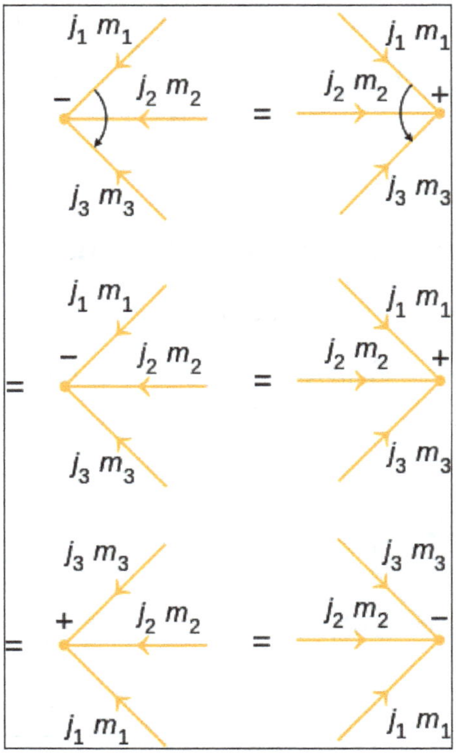

Time reversed equivalent.

For the inner product of two tensor product states:

$$\begin{aligned}&\left\langle j_n',m_n',\ldots,j_2',m_2',j_1',m_1' \mid j_1,m_1,j_2,m_2,\ldots j_n,m_n\right\rangle\\&=\langle j_n',m_n' \mid \ldots\langle j_2',m_2' \mid \langle j_1',m_1' \| j_1,m_1\rangle \mid j_2,m_2\rangle\ldots \mid j_n,m_n\rangle\\&=\prod_{k=1}^{n}\left\langle j_k',m_k' \mid j_k,m_k\right\rangle\end{aligned}$$

there are n lots of inner product arrows:

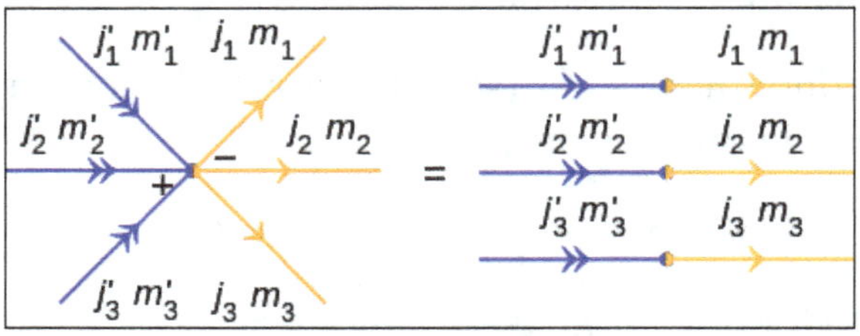

Inner product of $|j'_1, m'_1, j'_2, m'_2, j'_3, m'_3\rangle$ and $|j_1, m_1, j_2, m_2, j_3, m_3\rangle$, that is $\langle j'_3, m'_3, j'_2, m'_2, j'_1, m'_1 | j_1, m_1, j_2, m_2, j_3, m_3\rangle$.
Similarly for more than three pairs of angular momenta.

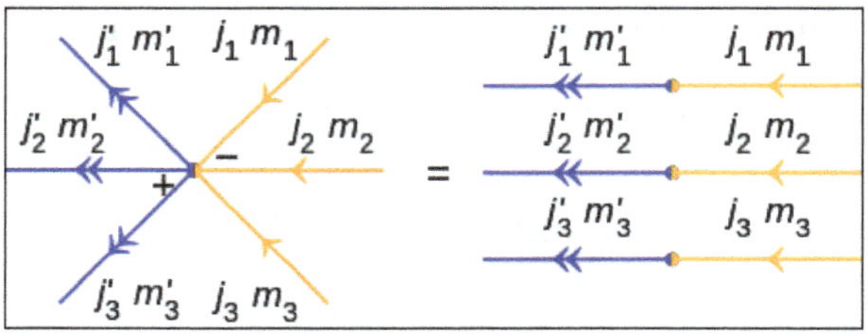

Time reversed equivalent.

Examples and Applications

- The diagrams are well-suited for Clebsch–Gordan coefficients.
- Calculations with real quantum systems, such as multielectron atoms and molecular systems.

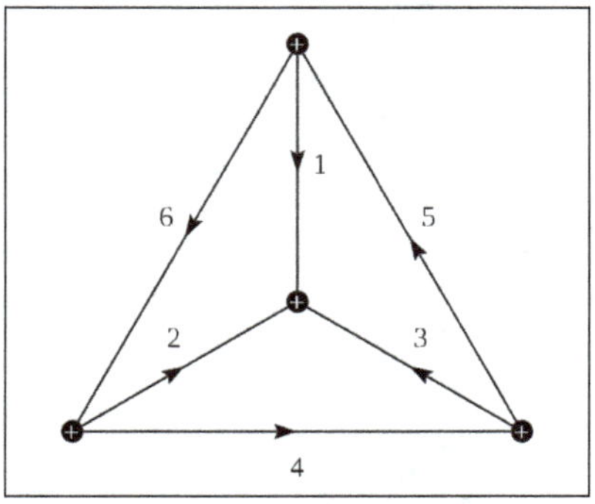

Diagram for a 6-j symbol, $\begin{Bmatrix} j_1 & j_2 & j_3 \\ j_4 & j_5 & j_6 \end{Bmatrix}$.

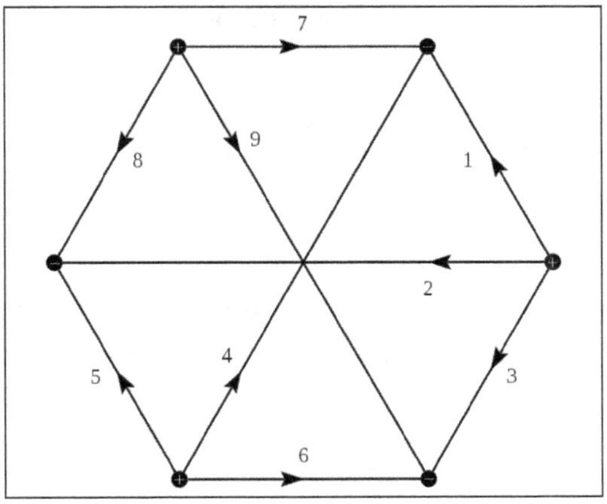

Diagram for a 9-j symbol, $\begin{Bmatrix} j_1 & j_2 & j_3 \\ j_4 & j_5 & j_6 \\ j_7 & j_8 & j_9 \end{Bmatrix}$.

ORBITAL ANGULAR MOMENTUM

Consider a particle described by the Cartesian coordinates $(x, y, z) \equiv \mathbf{x}$ and their conjugate momenta $(p_x, p_y, p_z) \equiv \mathbf{p}$. The classical definition of the orbital angular momentum of such a particle about the origin is $\mathbf{L} = \mathbf{x} \times \mathbf{p}$, giving:

$$L_x = y\,p_z - z\,p_y,$$
$$L_y = z\,p_x - x\,p_z,$$
$$L_z = x\,p_y - y\,p_x.$$

Let us assume that the operators $(L_x, L_y, L_z) \equiv \mathbf{L}$ which represent the components of orbital angular momentum in quantum mechanics can be defined in an analogous manner to the corresponding components of classical angular momentum. In other words, we are going to assume that the above equations specify the angular momentum operators in terms of the position and linear momentum operators. Note that L_x, L_y and L_z are Hermitian, so they represent things which can, in principle, be measured. Note, also, that there is no ambiguity regarding the order in which operators appear in products on the right-hand sides of equations $L_x = y\,p_z - z\,p_y$ to $L_z = x\,p_y - y\,p_x$. because all of the products consist of operators that commute.

The fundamental commutation relations satisfied by the position and linear momentum operators are:

$$[x_i, x_j] = 0,$$
$$[p_i, p_j] = 0,$$
$$[x_i, p_j] = \mathrm{i}\,\hbar\,\delta_{ij},$$

where i and j stand for either x, y, or z. Consider the commutator of the operators L_x and L_z:

$$[L_x, L_y] = [(y\,p_z - z\,p_y), (z\,p_x - x\,p_z)] = y[p_z, z]\,p_x + x\,p_y[z, p_z]$$

$$= \mathrm{i}\hbar(-y\,p_x + x\,p_y) = \mathrm{i}\hbar L_z.$$

The cyclic permutations of the above result yield the fundamental commutation relations satisfied by the components of an orbital angular momentum:

$$[L_x, L_y] = \mathrm{i}\hbar L_z,$$
$$[L_y, L_z] = \mathrm{i}\hbar L_x,$$
$$[L_z, L_x] = \mathrm{i}\hbar L_y.$$

These can be summed up more succinctly by writing:

$$\mathbf{L} \times \mathbf{L} = \mathrm{i}\hbar\mathbf{L}.$$

The three commutation equations $[L_x, L_y] = \mathrm{i}\hbar L_z$, $[L_y, L_z] = \mathrm{i}\hbar L_x$ and $[L_z, L_x] = \mathrm{i}\hbar L_y$ are the foundation for the whole theory of angular momentum in quantum mechanics. Whenever we encounter three operators having these commutation relations, we know that the dynamical variables that they represent have identical properties to those of the components of an angular momentum (which we are about to derive). In fact, we shall assume that any three operators that satisfy the commutation equations above represent the components of some sort of angular momentum.

Suppose that there are N particles in the system, with angular momentum vectors $\mathbf{L}_\mathrm{i}$ (where **i** runs from 1 to **N**). Each of these vectors satisfies equation $\mathbf{L} \times \mathbf{L} = \mathrm{i}\hbar\mathbf{L}$, so that:

$$\mathbf{L}_i \times \mathbf{L}_i = \mathrm{i}\hbar\mathbf{L}_i.$$

However, we expect the angular momentum operators belonging to different particles to commute, because they represent different degrees of freedom of the system. So, we can write:

$$\mathbf{L}_i \times \mathbf{L}_j + \mathbf{L}_j \times \mathbf{L}_i = 0,$$

for $i \neq j$. Consider the total angular momentum of the system, $\mathbf{L} = \sum_{i=1,N} \mathbf{L}_i$. It is clear from equations above that:

$$\mathbf{L} \times \mathbf{L} = \sum_{i=1,N} \mathbf{L}_i \times \sum_{j=1,N} \mathbf{L}_j = \sum_{i=1,N} L_i \times L_i + \frac{1}{2} \sum_{i,j=1,N}^{i\neq j} (\mathbf{L}_i \times \mathbf{L}_j + \mathbf{L}_j \times \mathbf{L}_i)$$
$$= \mathrm{i}\hbar \sum_{i=1,N} \mathbf{L}_i = \mathrm{i}\hbar\mathbf{L}.$$

Thus, the sum of two or more angular momentum vectors satisfies the same commutation relation as a primitive angular momentum vector. In particular, the total angular momentum of the system satisfies the commutation equation $\mathbf{L}_i \times \mathbf{L}_j + \mathbf{L}_j \times \mathbf{L}_i = 0$.

The immediate conclusion which can be drawn from the commutation equations $[L_x, L_y] = \mathrm{i}\hbar L_z$, $[L_y, L_z] = \mathrm{i}\hbar L_x$ and $[L_z, L_x] = \mathrm{i}\hbar L_y$ is that the three components of an angular momentum vector cannot be specified (or measured) simultaneously. In fact, once we have specified one component, the values of other two components become uncertain. It is conventional to specify the z-component, L_z.

Consider the magnitude squared of the angular momentum vector, $L^2 \equiv L_x^2 + L_y^2 + L_z^2$. The commutator of L^2 and L_z is written:

$$[L^2, L_z] = [L_x^2, L_z] + [L_y^2, L_z] + [L_z^2, L_z].$$

It is easily demonstrated that:

$$[L_x^2, L_z] = -\mathrm{i}\hbar (L_x L_y + L_y L_x),$$

$$[L_y^2, L_z] = +\mathrm{i}\hbar (L_x L_y + L_y L_x),$$

$$[L_z^2, L_z] = 0,$$

so

$$[L^2, L_z] = 0.$$

Because there is nothing special about the z-direction, we conclude that L^2 also commutes with L_x and L_y. It is clear from equations $[L_x, L_y] = \mathrm{i}\hbar L_z$, $[L_y, L_z] = \mathrm{i}\hbar L_x$ and $[L_z, L_x] = \mathrm{i}\hbar L_y$ that the best we can do in quantum mechanics is to specify the magnitude of an angular momentum vector along with one of its components (by convention, the z-component).

It is convenient to define the shift operators L^+ and L^-:

$$L^+ = L_x + \mathrm{i} L_y,$$

$$L^- = L_x - \mathrm{i} L_y.$$

It can easily be shown that:

$$[L^+, L_z] = -\hbar L^+,$$

$$[L^-, L_z] = +\hbar L^-,$$

$$[L^+, L^-] = 2\hbar L_z,$$

and also that both shift operators commute with L^2.

SPIN

In quantum mechanics and particle physics, spin is an intrinsic form of angular momentum carried by elementary particles, composite particles (hadrons), and atomic nuclei. Spin is one of two types of angular momentum in quantum mechanics, the other being orbital angular momentum. The orbital angular momentum operator is the quantum-mechanical counterpart to the classical angular momentum of orbital revolution and appears when there is periodic structure to its wavefunction as the angle varies. The existence of spin angular momentum is inferred from experiments, such as the Stern–Gerlach experiment, in which silver atoms were observed to possess two possible discrete angular momenta despite having no orbital angular momentum.

In some ways, spin is like a vector quantity; it has a definite magnitude, and it has a "direction" (but quantization makes this "direction" different from the direction of an ordinary vector). All elementary particles of a given kind have the same magnitude of spin angular momentum, which is indicated by assigning the particle a spin quantum number.

The SI unit of spin is the (N·m·s) or ($kg·m^2·s^{-1}$), just as with classical angular momentum. In practice, spin is given as a dimensionless spin quantum number by dividing the spin angular momentum by the reduced Planck constant $\hbar$, which has the same units of angular momentum, although this is not the full computation of this value. Very often, the "spin quantum number" is simply called "spin", The fact that it is a quantum number is implicit. When combined with the spin-statistics theorem, the spin of electrons results in the Pauli exclusion principle, which in turn underlies the periodic table of chemical elements.

Wolfgang Pauli in 1924 was the first to propose a doubling of electron states due to a two-valued non-classical "hidden rotation". In 1925, George Uhlenbeck and Samuel Goudsmit at Leiden University suggested the simple physical interpretation of a particle spinning around its own axis, in the spirit of the old quantum theory of Bohr and Sommerfeld. Ralph Kronig anticipated the Uhlenbeck-Goudsmit model in discussion with Hendrik Kramers several months earlier in Copenhagen, but did not publish. The mathematical theory was worked out in depth by Pauli in 1927. When Paul Dirac derived his relativistic quantum mechanics in 1928, electron spin was an essential part of it.

Quantum Number

As the name suggests, spin was originally conceived as the rotation of a particle around some axis. This picture is correct so far as spin obeys the same mathematical laws as quantized angular momenta do. On the other hand, spin has some peculiar properties that distinguish it from orbital angular momenta:

- Spin quantum numbers may take half-integer values.

- Although the direction of its spin can be changed, an elementary particle cannot be made to spin faster or slower.
- The spin of a charged particle is associated with a magnetic dipole moment with a *g*-factor differing from 1. This could only occur classically if the internal charge of the particle were distributed differently from its mass.

The conventional definition of the spin quantum number, *s*, is $s = n/2$, where *n* can be any non-negative integer. Hence the allowed values of *s* are 0, 1/2, 1, 3/2, 2, etc. The value of *s* for an elementary particle depends only on the type of particle, and cannot be altered in any known way. The spin angular momentum, *S*, of any physical system is quantized. The allowed values of *S* are:

$$S = \hbar\sqrt{s(s+1)} = \frac{h}{4\pi}\sqrt{n(n+2)},$$

where *h* is the Planck constant and $\hbar = \frac{h}{2\pi}$ is the reduced Planck constant. In contrast, orbital angular momentum can only take on integer values of *s*; i.e., even-numbered values of *n*.

Fermions and Bosons

Those particles with half-integer spins, such as $\frac{1}{2}, \frac{3}{2}, \frac{5}{2}$, are known as fermions, while those particles with integer spins, such as 0, 1, 2, are known as bosons. The two families of particles obey different rules and *broadly* have different roles in the world around us. A key distinction between the two families is that fermions obey the Pauli exclusion principle; that is, there cannot be two identical fermions simultaneously having the same quantum numbers (meaning, roughly, having the same position, velocity and spin direction). In contrast, bosons obey the rules of Bose–Einstein statistics and have no such restriction, so they may "bunch together" even if in identical states. Also, composite particles can have spins different from their component particles. For example, a helium atom in the ground state has spin 0 and behaves like a boson, even though the quarks and electrons which make it up are all fermions.

This has some profound consequences:

- Quarks and leptons (including electrons and neutrinos), which make up what is classically known as matter, are all fermions with spin $\frac{1}{2}$. The common idea that "matter takes up space" actually comes from the Pauli exclusion principle acting on these particles to prevent the fermions that make up matter from being in the same quantum state. Further compaction would require electrons to occupy the same energy states, and therefore a kind of pressure (sometimes known as degeneracy pressure of electrons) acts to resist the fermions being overly close.

Elementary fermions with other spins $\left(\frac{3}{2}, \frac{5}{2}, \text{etc.}\right)$ are not known to exist.

- Elementary particles which are thought of as carrying forces are all bosons with spin 1. They include the photon which carries the electromagnetic force, the gluon (strong force), and the W and Z bosons (weak force). The ability of bosons to occupy the same quantum state is used in the laser, which aligns many photons having the same quantum number (the same direction and frequency), superfluid liquid helium resulting from helium-4 atoms being bosons, and superconductivity where pairs of electrons (which individually are fermions) act as single composite bosons.

  Elementary bosons with other spins (0, 2, 3, etc.) were not historically known to exist, although they have received considerable theoretical treatment and are well established within their respective mainstream theories. In particular, theoreticians have proposed the graviton (predicted to exist by some quantum gravity theories) with spin 2, and the Higgs boson (explaining electroweak symmetry breaking) with spin 0. Since 2013, the Higgs boson with spin 0 has been considered proven to exist. It is the first scalar elementary particle (spin 0) known to exist in nature.

Spin-statistics Theorem

The spin-statistics theorem states that particles with half-integer spin (fermions) obey Fermi–Dirac statistics and the Pauli Exclusion Principle, and that particles with integer spin (bosons) obey Bose–Einstein statistics, occupy "symmetric states", and thus can share quantum states. The theorem relies on both quantum mechanics and the theory of special relativity, and this connection between spin and statistics has been called "one of the most important applications of the special relativity theory".

Magnetic Moments

Particles with spin can possess a magnetic dipole moment, just like a rotating electrically charged body in classical electrodynamics. These magnetic moments can be experimentally observed in several ways, e.g. by the deflection of particles by inhomogeneous magnetic fields in a Stern–Gerlach experiment, or by measuring the magnetic fields generated by the particles themselves.

The intrinsic magnetic moment $\boldsymbol{\mu}$ of a spin $\frac{1}{2}$ particle with charge q, mass m, and spin angular momentum $\mathbf{S}$, is:

$$\mu = \frac{g_s q}{2m}\mathbf{S}$$

where the dimensionless quantity g_s is called the spin g-factor. For exclusively orbital rotations it would be 1 (assuming that the mass and the charge occupy spheres of equal radius).

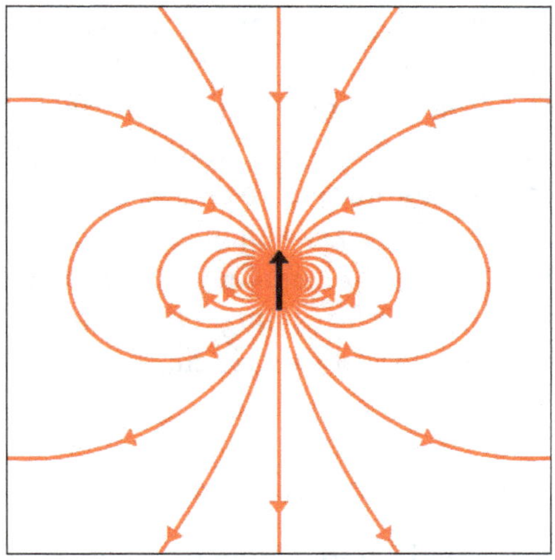

Schematic diagram depicting the spin of the neutron as the black arrow and magnetic field lines associated with the neutron magnetic moment. The neutron has a negative magnetic moment. While the spin of the neutron is upward in this diagram, the magnetic field lines at the center of the dipole are downward.

The electron, being a charged elementary particle, possesses a nonzero magnetic moment. One of the triumphs of the theory of quantum electrodynamics is its accurate prediction of the electron g-factor, which has been experimentally determined to have the value −2.00231930436256, with the digits in parentheses denoting measurement uncertainty in the last two digits at one standard deviation. The value of 2 arises from the Dirac equation, a fundamental equation connecting the electron's spin with its electromagnetic properties, and the correction of 0.002319304... arises from the electron's interaction with the surrounding electromagnetic field, including its own field.

Composite particles also possess magnetic moments associated with their spin. In particular, the neutron possesses a non-zero magnetic moment despite being electrically neutral. This fact was an early indication that the neutron is not an elementary particle. In fact, it is made up of quarks, which are electrically charged particles. The magnetic moment of the neutron comes from the spins of the individual quarks and their orbital motions.

Neutrinos are both elementary and electrically neutral. The minimally extended Standard Model that takes into account non-zero neutrino masses predicts neutrino magnetic moments of:

$$\mu_\nu \approx 3\times10^{-19}\,\mu_{\mathrm{B}}\frac{m_\nu}{\mathrm{eV}}$$

where the μ_ν are the neutrino magnetic moments, m_ν are the neutrino masses, and μ_{B} is the Bohr magneton. New physics above the electroweak scale could, however, lead to significantly higher neutrino magnetic moments. It can be shown in a model independent way that neutrino magnetic moments larger than about $10^{-14}\ \mu_B$ are "unnatural" because they would also lead to large radiative contributions to the neutrino mass. Since the neutrino masses are known to be at most about 1 eV, the large radiative corrections would then have to be "fine-tuned" to cancel each other, to a large degree, and leave the neutrino mass small. The measurement of neutrino magnetic moments is an active area of research. Experimental results have put the neutrino magnetic moment at less than 1.2×10^{-10} times the electron's magnetic moment. On the other hand elementary particles with spin but without electric charge, such as a photon or a Z boson, do not have a magnetic moment.

Curie Temperature and Loss of Alignment

In ordinary materials, the magnetic dipole moments of individual atoms produce magnetic fields that cancel one another, because each dipole points in a random direction, with the overall average being very near zero. Ferromagnetic materials below their Curie temperature, however, exhibit magnetic domains in which the atomic dipole moments are locally aligned, producing a macroscopic, non-zero magnetic field from the domain. These are the ordinary "magnets" with which we are all familiar.

In paramagnetic materials, the magnetic dipole moments of individual atoms spontaneously align with an externally applied magnetic field. In diamagnetic materials, on the other hand, the magnetic dipole moments of individual atoms spontaneously align oppositely to any externally applied magnetic field, even if it requires energy to do so.

The study of the behavior of such "spin models" is a thriving area of research in condensed matter physics. For instance, the Ising model describes spins (dipoles) that have only two possible states, up and down, whereas in the Heisenberg model the spin vector is allowed to point in any direction. These models have many interesting properties, which have led to interesting results in the theory of phase transitions.

Direction

Spin Projection Quantum Number and Multiplicity

In classical mechanics, the angular momentum of a particle possesses not only a magnitude (how fast the body is rotating), but also a direction (either up or down on the axis of rotation of the particle). Quantum mechanical spin also contains information about direction, but in a more subtle form. Quantum mechanics states that the component of angular momentum measured along any direction can only take on the values:

$$S_i = \hbar s_i, \quad s_i \in \{-s, -(s-1), \ldots, s-1, s\}$$

where S_i is the spin component along the i-axis (either x, y, or z), s_i is the spin projection quantum number along the i-axis, and s is the principal spin quantum number. Conventionally the direction chosen is the z-axis:

$$S_z = \hbar s_z, \quad s_z \in \{-s, -(s-1), \ldots, s-1, s\}$$

where S_z is the spin component along the z-axis, s_z is the spin projection quantum number along the z-axis.

One can see that there are 2s + 1 possible values of s_z. The number "2s + 1" is the multiplicity of the spin system. For example, there are only two possible values for a spin-1/2 particle: $s_z = +\frac{1}{2}$ and $s_z = -\frac{1}{2}$. These correspond to quantum states in which the spin component is pointing in the +z or −z directions respectively, and are often referred to as "spin up" and "spin down". For a spin-$\frac{3}{2}$ particle, like a delta baryon, the possible values are $+\frac{3}{2}, +\frac{1}{2}, -\frac{1}{2}, -\frac{3}{2}$.

Vector

For a given quantum state, one could think of a spin vector $\langle S \rangle$ whose components are the expectation values of the spin components along each axis, i.e., $\langle S \rangle = [\langle S_x \rangle, \langle S_y \rangle, \langle S_z \rangle]$ This vector then would describe the "direction" in which the spin is pointing, corresponding to the classical concept of the axis of rotation. It turns out that the spin vector is not very useful in actual quantum mechanical calculations, because it cannot be measured directly: s_x, s_y and s_z cannot possess simultaneous definite values, because of a quantum uncertainty relation between them. However, for statistically large collections of particles that have been placed in the same pure quantum state, such as through the use of a Stern–Gerlach apparatus, the spin vector does have a well-defined experimental meaning: It specifies the direction in ordinary space in which a subsequent detector must be oriented in order to achieve the maximum possible probability (100%) of detecting every particle in the collection. For spin-$\frac{1}{2}$ particles, this maximum probability drops off smoothly as the angle between the spin vector and the detector increases, until at an angle of 180 degrees—that is, for detectors oriented in the opposite direction to the spin vector. The expectation of detecting particles from the collection reaches a minimum of 0%.

As a qualitative concept, the spin vector is often handy because it is easy to picture classically. For instance, quantum mechanical spin can exhibit phenomena analogous to classical gyroscopic effects. For example, one can exert a kind of "torque" on an electron by putting it in a magnetic field (the field acts upon the electron's intrinsic magnetic dipole moment). The result is that the spin vector undergoes precession, just like a classical gyroscope. This phenomenon is known as electron spin resonance (ESR). The equivalent behaviour of protons in atomic nuclei is used in nuclear magnetic resonance (NMR) spectroscopy and imaging.

Mathematically, quantum-mechanical spin states are described by vector-like objects known as spinors. There are subtle differences between the behavior of spinors and vectors under coordinate rotations. For example, rotating a spin-$\frac{1}{2}$ particle by 360 degrees does not bring it back to the same quantum state, but to the state with the opposite quantum phase; this is detectable, in principle, with interference experiments. To return the particle to its exact original state, one needs a 720-degree rotation. (The Plate trick and Möbius strip give non-quantum analogies). A spin-zero particle can only have a single quantum state, even after torque is applied. Rotating a spin-2 particle 180 degrees can bring it back to the same quantum state and a spin-4 particle should be rotated 90 degrees to bring it back to the same quantum state. The spin-2 particle can be analogous to a straight stick that looks the same even after it is rotated 180 degrees and a spin 0 particle can be imagined as sphere, which looks the same after whatever angle it is turned through.

Mathematical Formulation

Operator

Spin obeys commutation relations analogous to those of the orbital angular momentum:

$$[S_j, S_k] = i\hbar \varepsilon_{jkl} S_l$$

where ε_{jkl} is the Levi-Civita symbol. It follows (as with angular momentum) that the eigenvectors of S^2 and S_z (expressed as kets in the total S basis) are:

$$S^2 \mid s, m_s\rangle = \hbar^2 s(s+1) \mid s, m_s\rangle$$

$$S_z \mid s, m_s\rangle = \hbar m_s \mid s, m_s\rangle$$

The spin raising and lowering operators acting on these eigenvectors give:

$$S_\pm \mid s, m_s\rangle = \hbar\sqrt{s(s+1) - m_s(m_s \pm 1)} \mid s, m_s \pm 1\rangle$$

where $S_\pm = S_x \pm i\, S_y$.

But unlike orbital angular momentum the eigenvectors are not spherical harmonics. They are not functions of θ and φ. There is also no reason to exclude half-integer values of s and m_s.

In addition to their other properties, all quantum mechanical particles possess an intrinsic spin (though this value may be equal to zero). The spin is quantized in units of the reduced Planck constant, such that the state function of the particle is, say, not $\psi = \psi(\mathbf{r})$, but $\psi = \psi(\mathbf{r},\sigma)$ where σ is out of the following discrete set of values:

$$\sigma \in \{-s\hbar, -(s-1)\hbar, \cdots, +(s-1)\hbar, +s\hbar\}$$

One distinguishes bosons (integer spin) and fermions (half-integer spin). The total angular momentum conserved in interaction processes is then the sum of the orbital angular momentum and the spin.

Pauli Matrices

The quantum mechanical operators associated with spin-$\frac{1}{2}$ observables are:

$$\hat{\mathbf{S}} = \frac{\hbar}{2}\sigma$$

where in Cartesian components:

$$S_x = \frac{\hbar}{2}\sigma_x, \quad S_y = \frac{\hbar}{2}\sigma_y, \quad S_z = \frac{\hbar}{2}\sigma_z .$$

For the special case of spin-$\frac{1}{2}$ particles, σ_x, σ_y and σ_z are the three Pauli matrices, given by:

$$\sigma_x = \begin{pmatrix} 0 & 1 \\ 1 & 0 \end{pmatrix} \quad \sigma_y = \begin{pmatrix} 0 & -i \\ i & 0 \end{pmatrix} \quad \sigma_z = \begin{pmatrix} 1 & 0 \\ 0 & -1 \end{pmatrix}.$$

Pauli Exclusion Principle

For systems of N identical particles this is related to the Pauli exclusion principle, which states that by interchanges of any two of the N particles one must have:

$$\psi(\cdots \mathbf{r}_i, \sigma_i \cdots \mathbf{r}_j, \sigma_j \cdots) = (-1)^{2s}\psi(\cdots \mathbf{r}_j, \sigma_j \cdots \mathbf{r}_i, \sigma_i \cdots)$$

Thus, for bosons the prefactor $(-1)^{2s}$ will reduce to +1, for fermions to −1. In quantum mechanics all particles are either bosons or fermions. In some speculative relativistic quantum field theories "supersymmetric" particles also exist, where linear combinations of bosonic and fermionic components appear. In two dimensions, the prefactor $(-1)^{2s}$ can be replaced by any complex number of magnitude 1 such as in the anyon. The above permutation postulate for N-particle state functions has most-important consequences in daily life, e.g. the periodic table of the chemical elements.

Rotations

Quantum mechanics states that components of angular momentum measured along any direction can only take a number of discrete values. The most convenient quantum mechanical description of particle's spin is therefore with a set of complex numbers corresponding to amplitudes of finding a given value of projection of its intrinsic angular momentum on a given axis. For instance, for a spin-$\frac{1}{2}$ particle, we would need two numbers $a_{\pm\frac{1}{2}}$, giving amplitudes of finding it with projection of angular momentum equal to $\frac{\hbar}{2}$ and $-\frac{\hbar}{2}$, satisfying the requirement:

$$\left|a_{\frac{1}{2}}\right|^2 + \left|a_{-\frac{1}{2}}\right|^2 = 1.$$

For a generic particle with spin *s*, we would need 2s + 1 such parameters. Since these numbers depend on the choice of the axis, they transform into each other non-trivially when this axis is rotated. It's clear that the transformation law must be linear, so we can represent it by associating a matrix with each rotation, and the product of two transformation matrices corresponding to rotations A and B must be equal (up to phase) to the matrix representing rotation AB. Further, rotations preserve the quantum mechanical inner product, and so should our transformation matrices:

$$\sum_{m=-j}^{j} a_m^* b_m = \sum_{m=-j}^{j} \left(\sum_{n=-j}^{j} U_{nm} a_n \right)^* \left(\sum_{k=-j}^{j} U_{km} b_k \right)$$

$$\sum_{n=-j}^{j} \sum_{k=-j}^{j} U_{np}^* U_{kq} = \delta_{pq}.$$

Mathematically speaking, these matrices furnish a unitary projective representation of the rotation group SO(3). Each such representation corresponds to a representation of the covering group of SO(3), which is SU(2). There is one *n*-dimensional irreducible representation of SU(2) for each dimension, though this representation is *n*-dimensional real for odd *n* and *n*-dimensional complex for even *n* (hence of real dimension 2*n*). For a rotation by angle θ in the plane with normal vector $\hat{}$, *U* can be written:

$$U = e^{-\frac{i}{\hbar}\theta\cdot\mathbf{S}},$$

where $\theta = \theta\hat{\theta}$ and **S** is the vector of spin operators.

A generic rotation in 3-dimensional space can be built by compounding operators of this type using Euler angles:

$$\mathcal{R}(\alpha,\beta,\gamma) = e^{-i\alpha S_z} e^{-i\beta S_y} e^{-i\gamma S_z}$$

An irreducible representation of this group of operators is furnished by the Wigner D-matrix:

$$D^s_{m'm}(\alpha,\beta,\gamma) \equiv \langle sm' | \mathcal{R}(\alpha,\beta,\gamma) | sm \rangle = e^{-im'\alpha} d^s_{m'm}(\beta) e^{-im\gamma},$$

where,

$$d^s_{m'm}(\beta) = \langle sm' | e^{-i\beta s_y} | sm \rangle$$

is Wigner's small d-matrix. Note that for $\gamma = 2\pi$ and $\alpha = \beta = 0$; i.e., a full rotation about the z-axis, the Wigner D-matrix elements become:

$$D^{s}_{m'm}(0,0,2\pi) = d^{s}_{m'm}(0)e^{-im2\pi} = \delta_{m'm}(-1)^{2m}.$$

Recalling that a generic spin state can be written as a superposition of states with definite m, we see that if s is an integer, the values of m are all integers, and this matrix corresponds to the identity operator. However, if s is a half-integer, the values of m are also all half-integers, giving $(-1)^{2m} = -1$ for all m, and hence upon rotation by 2π the state picks up a minus sign. This fact is a crucial element of the proof of the spin-statistics theorem.

Lorentz Transformations

We could try the same approach to determine the behavior of spin under general Lorentz transformations, but we would immediately discover a major obstacle. Unlike SO(3), the group of Lorentz transformations SO(3,1) is non-compact and therefore does not have any faithful, unitary, finite-dimensional representations.

In case of spin-1/2 particles, it is possible to find a construction that includes both a finite-dimensional representation and a scalar product that is preserved by this representation. We associate a 4-component Dirac spinor ψ with each particle. These spinors transform under Lorentz transformations according to the law:

$$\psi' = \exp\left(\tfrac{1}{8}\omega_{\mu\nu}[\gamma_{\mu},\gamma_{\nu}]\right)\psi$$

where γ_ν are gamma matrices and $\omega_{\mu\nu}$ is an antisymmetric 4×4 matrix parametrizing the transformation. It can be shown that the scalar product:

$$\langle\psi \mid \phi\rangle = \bar{\psi}\phi = \psi^{\dagger}\gamma_0\phi$$

is preserved. It is not, however, positive definite, so the representation is not unitary.

Measurement of Spin Along the *x*-, *y*-, or *z*-axes

Each of the (Hermitian) Pauli matrices has two eigenvalues, +1 and −1. The corresponding normalized eigenvectors are:

$$\psi_{x+} = \left|\frac{1}{2},\frac{+1}{2}\right\rangle_x = \frac{1}{\sqrt{2}}\begin{pmatrix}1\\1\end{pmatrix}, \quad \psi_{x-} = \left|\frac{1}{2},\frac{-1}{2}\right\rangle_x = \frac{1}{\sqrt{2}}\begin{pmatrix}-1\\1\end{pmatrix},$$

$$\psi_{y+} = \left|\frac{1}{2},\frac{+1}{2}\right\rangle_y = \frac{1}{\sqrt{2}}\begin{pmatrix}1\\i\end{pmatrix}, \quad \psi_{y-} = \left|\frac{1}{2},\frac{-1}{2}\right\rangle_y = \frac{1}{\sqrt{2}}\begin{pmatrix}1\\-i\end{pmatrix},$$

$$\psi_{z+} = \left|\frac{1}{2},\frac{+1}{2}\right\rangle_z = \begin{pmatrix}1\\0\end{pmatrix}, \quad \psi_{z-} = \left|\frac{1}{2},\frac{-1}{2}\right\rangle_z = \begin{pmatrix}0\\1\end{pmatrix}.$$

Because any eigenvector multiplied by a constant is still an eigenvector, there is ambiguity about the overall sign. the convention is chosen to make the first element imaginary and negative if there is a sign ambiguity.

By the postulates of quantum mechanics, an experiment designed to measure the electron spin on the *x*-, *y*-, or *z*-axis can only yield an eigenvalue of the corresponding spin operator (S_x, S_y or S_z) on that axis, i.e. $\frac{\hbar}{2}$ or $-\frac{\hbar}{2}$. The quantum state of a particle (with respect to spin), can be represented by a two component spinor:

$$\psi = \begin{pmatrix} a+bi \\ c+di \end{pmatrix}.$$

When the spin of this particle is measured with respect to a given axis (in this example, the *x*-axis), the probability that its spin will be measured as $\frac{\hbar}{2}$ is just $|\langle\psi_{x+}|\psi\rangle|^2$. Correspondingly, the probability that its spin will be measured as $-\frac{\hbar}{2}$ is just $|\langle\psi_{x-}|\psi\rangle|^2$. Following the measurement, the spin state of the particle will collapse into the corresponding eigenstate. As a result, if the particle's spin along a given axis has been measured to have a given eigenvalue, all measurements will yield the same eigenvalue (since $|\langle\psi_{x+}|\psi_{x+}\rangle|^2 = 1$, etc), provided that no measurements of the spin are made along other axes.

Measurement of Spin along an Arbitrary Axis

The operator to measure spin along an arbitrary axis direction is easily obtained from the Pauli spin matrices. Let $u = (u_x, u_y, u_z)$ be an arbitrary unit vector. Then the operator for spin in this direction is simply:

$$S_u = \frac{\hbar}{2}(u_x\sigma_x + u_y\sigma_y + u_z\sigma_z).$$

The operator S_u has eigenvalues of $\pm\frac{\hbar}{2}$, just like the usual spin matrices. This method of finding the operator for spin in an arbitrary direction generalizes to higher spin states, one takes the dot product of the direction with a vector of the three operators for the three *x*-, *y*-, *z*-axis directions.

A normalized spinor for spin-$\frac{1}{2}$ in the (u_x, u_y, u_z) direction (which works for all spin states except spin down where it will give $\frac{0}{0}$), is:

$$\frac{1}{\sqrt{2+2u_z}}\begin{pmatrix} 1+u_z \\ u_x+iu_y \end{pmatrix}.$$

The above spinor is obtained in the usual way by diagonalizing the σ_u matrix and finding the eigenstates corresponding to the eigenvalues. In quantum mechanics, vectors are termed "normalized" when multiplied by a normalizing factor, which results in the vector having a length of unity.

Compatibility of Spin Measurements

Since the Pauli matrices do not commute, measurements of spin along the different axes are incompatible. This means that if, for example, we know the spin along the x-axis, and we then measure the spin along the y-axis, we have invalidated our previous knowledge of the x-axis spin. This can be seen from the property of the eigenvectors (i.e. eigenstates) of the Pauli matrices that:

$$\left|\langle \psi_{x\pm} | \psi_{y\pm} \rangle\right|^2 = \left|\langle \psi_{x\pm} | \psi_{z\pm} \rangle\right|^2 = \left|\langle \psi_{y\pm} | \psi_{z\pm} \rangle\right|^2 = \frac{1}{2}.$$

So when physicists measure the spin of a particle along the x-axis as, for example, $\frac{\hbar}{2}$, the particle's spin state collapses into the eigenstate $|\psi_{x+}\rangle$. When we then subsequently measure the particle's spin along the y-axis, the spin state will now collapse into either $|\psi_{y+}\rangle$ or $||\psi_{y-}\rangle$, each with probability $\frac{1}{2}$. Let us say, in our example, that we measure $-\frac{\hbar}{2}$. When we now return to measure the particle's spin along the x-axis again, the probabilities that we will measure $\frac{\hbar}{2}$ or $-\frac{\hbar}{2}$ are each $\frac{1}{2}$ (i.e. they are $\left|\langle \psi_{x+} | \psi_{y-} \rangle\right|^2$ and $\left|\langle \psi_{x-} | \psi_{y-} \rangle\right|^2$ respectively). This implies that the original measurement of the spin along the x-axis is no longer valid, since the spin along the x-axis will now be measured to have either eigenvalue with equal probability.

Higher Spins

The spin-$\frac{1}{2}$ operator $\mathbf{S} = \frac{\hbar}{2}\boldsymbol{\sigma}$ forms the fundamental representation of SU(2). By taking Kronecker products of this representation with itself repeatedly, one may construct all higher irreducible representations. That is, the resulting spin operators for higher spin systems in three spatial dimensions, for arbitrarily large s, can be calculated using this spin operator and ladder operators.

The resulting spin matrices and eigenvalues in the z-basis:

- For spin 1 they are:

$$= \frac{1}{\sqrt{2}}\begin{pmatrix} 0 & 1 & 0 \\ 1 & 0 & 1 \\ 0 & 1 & 0 \end{pmatrix}, \quad |1,+1\rangle_x = \frac{1}{2}\begin{pmatrix} 1 \\ \sqrt{2} \\ 1 \end{pmatrix}, \quad |1,0\rangle_x = \frac{1}{\sqrt{2}}\begin{pmatrix} 1 \\ 0 \\ 1 \end{pmatrix}, \quad |1,-1\rangle_x = \frac{1}{2}\begin{pmatrix} 1 \\ -\sqrt{2} \\ 1 \end{pmatrix}$$

$$S_y = \frac{\hbar}{\sqrt{2}}\begin{pmatrix} 0 & -i & 0 \\ i & 0 & -i \\ 0 & i & 0 \end{pmatrix},\ |1,+1\rangle_y = \frac{1}{2}\begin{pmatrix} -1 \\ -i\sqrt{2} \\ 1 \end{pmatrix},\ |1,0\rangle_y = \frac{1}{\sqrt{2}}\begin{pmatrix} 1 \\ 0 \\ 1 \end{pmatrix},\ |1,-1\rangle_y = \frac{1}{2}\begin{pmatrix} -1 \\ i\sqrt{2} \\ 1 \end{pmatrix}$$

$$S_z = \hbar\begin{pmatrix} 1 & 0 & 0 \\ 0 & 0 & 0 \\ 0 & 0 & -1 \end{pmatrix},\ |1,+1\rangle_z = \begin{pmatrix} 1 \\ 0 \\ 0 \end{pmatrix},\ |1,0\rangle_z = \begin{pmatrix} 0 \\ 1 \\ 0 \end{pmatrix},\ |1,-1\rangle_z = \begin{pmatrix} 0 \\ 0 \\ 1 \end{pmatrix}$$

- For spin $\frac{3}{2}$ they are:

$$S_x = \frac{\hbar}{2}\begin{pmatrix} 0 & \sqrt{3} & 0 & 0 \\ \sqrt{3} & 0 & 2 & 0 \\ 0 & 2 & 0 & \sqrt{3} \\ 0 & 0 & \sqrt{3} & 0 \end{pmatrix},\left|\frac{3}{2},\frac{+3}{2}\right\rangle_x = \frac{1}{2\sqrt{2}}\begin{pmatrix} 1 \\ \sqrt{3} \\ \sqrt{3} \\ 1 \end{pmatrix},\left|\frac{3}{2},\frac{+1}{2}\right\rangle_x$$

$$= \frac{1}{2\sqrt{2}}\begin{pmatrix} -\sqrt{3} \\ -1 \\ 1 \\ \sqrt{3} \end{pmatrix},\left|\frac{3}{2},\frac{-1}{2}\right\rangle_x = \frac{1}{2\sqrt{2}}\begin{pmatrix} \sqrt{3} \\ -1 \\ -1 \\ \sqrt{3} \end{pmatrix},\left|\frac{3}{2},\frac{-3}{2}\right\rangle_x = \frac{1}{2\sqrt{2}}\begin{pmatrix} -1 \\ \sqrt{3} \\ -\sqrt{3} \\ 1 \end{pmatrix}$$

$$S_y = \frac{\hbar}{2}\begin{pmatrix} 0 & -i\sqrt{3} & 0 & 0 \\ i\sqrt{3} & 0 & -2i & 0 \\ 0 & 2i & 0 & -i\sqrt{3} \\ 0 & 0 & i\sqrt{3} & 0 \end{pmatrix},\left|\frac{3}{2},\frac{+3}{2}\right\rangle_y = \frac{1}{2\sqrt{2}}\begin{pmatrix} i \\ -\sqrt{3} \\ -i\sqrt{3} \\ 1 \end{pmatrix},\left|\frac{3}{2},\frac{+1}{2}\right\rangle_y$$

$$= \frac{1}{2\sqrt{2}}\begin{pmatrix} -i\sqrt{3} \\ 1 \\ -i \\ \sqrt{3} \end{pmatrix},\left|\frac{3}{2},\frac{-1}{2}\right\rangle_y = \frac{1}{2\sqrt{2}}\begin{pmatrix} i\sqrt{3} \\ 1 \\ i \\ \sqrt{3} \end{pmatrix},\left|\frac{3}{2},\frac{-3}{2}\right\rangle_y = \frac{1}{2\sqrt{2}}\begin{pmatrix} -i \\ -\sqrt{3} \\ i\sqrt{3} \\ 1 \end{pmatrix}$$

$$S_z = \frac{\hbar}{2}\begin{pmatrix} 3 & 0 & 0 & 0 \\ 0 & 1 & 0 & 0 \\ 0 & 0 & -1 & 0 \\ 0 & 0 & 0 & -3 \end{pmatrix},\ \left|\frac{3}{2},\frac{+3}{2}\right\rangle_z = \begin{pmatrix} 1 \\ 0 \\ 0 \\ 0 \end{pmatrix},\ \left|\frac{3}{2},\frac{+1}{2}\right\rangle_z$$

$$= \begin{pmatrix} 0 \\ 0 \\ 1 \\ 0 \end{pmatrix},\ \left|\frac{3}{2},\frac{-1}{2}\right\rangle_z = \begin{pmatrix} 0 \\ 0 \\ 1 \\ 0 \end{pmatrix},\ \left|\frac{3}{2},\frac{-3}{2}\right\rangle_z = \begin{pmatrix} 0 \\ 0 \\ 0 \\ 1 \end{pmatrix}$$

- The generalization of these matrices for arbitrary spin s is:

$$\left(S_x\right)_{ab} = \frac{\hbar}{2}\left(\delta_{a,b+1} + \delta_{a+1,b}\right)\sqrt{(s+1)(a+b-1)-ab}$$

$$\left(S_y\right)_{ab} = \frac{i\hbar}{2}\left(\delta_{a,b+1} - \delta_{a+1,b}\right)\sqrt{(s+1)(a+b-1)-ab}$$

$$\left(S_z\right)_{ab} = \hbar(s+1-a)\delta_{a,b} = \hbar(s+1-b)\delta_{a,b}$$

where indices a, b are integer numbers such that:

$$1 \le a \le 2s+1 \text{ and}$$
$$1 \le b \le 2s+1$$

Also useful in the quantum mechanics of multiparticle systems, the general Pauli group G_n is defined to consist of all *n*-fold tensor products of Pauli matrices.

The analog formula of Euler's formula in terms of the Pauli matrices:

$$e^{i\theta(\hat{\mathbf{n}}\cdot\sigma)} = I\cos(\theta) + i\left(\hat{\mathbf{n}}\cdot\sigma\right)\sin(\theta)$$

for higher spins is tractable, but less simple.

Parity

In tables of the spin quantum number *s* for nuclei or particles, the spin is often followed by a "+" or "−". This refers to the parity with "+" for even parity (wave function unchanged by spatial inversion) and "−" for odd parity (wave function negated by spatial inversion). For Bi-209, the only stable isotope, the entry 9/2− means that the nuclear spin is 9/2 and the parity is odd.

Applications

Spin has important theoretical implications and practical applications. Well-established *direct* applications of spin include:

- Nuclear magnetic resonance (NMR) spectroscopy in chemistry.
- Electron spin resonance spectroscopy in chemistry and physics.
- Magnetic resonance imaging (MRI) in medicine, a type of applied NMR, which relies on proton spin density.
- Giant magnetoresistive (GMR) drive head technology in modern hard disks.

Electron spin plays an important role in magnetism, with applications for instance in computer memories. The manipulation of *nuclear spin* by radiofrequency waves (nuclear magnetic resonance) is important in chemical spectroscopy and medical imaging.

Spin-orbit coupling leads to the fine structure of atomic spectra, which is used in atomic clocks and in the modern definition of the second. Precise measurements of the g-factor of the electron have played an important role in the development and verification of quantum electrodynamics. *Photon spin* is associated with the polarization of light.

An emerging application of spin is as a binary information carrier in spin transistors. The original concept, proposed in 1990, is known as Datta-Das spin transistor. Electronics based on spin transistors are referred to as spintronics. The manipulation of spin in dilute magnetic semiconductor materials, such as metal-doped ZnO or TiO_2 imparts a further degree of freedom and has the potential to facilitate the fabrication of more efficient electronics.

There are many *indirect* applications and manifestations of spin and the associated Pauli exclusion principle, starting with the periodic table of chemistry.

THE STERN-GERLACH EXPERIMENT

The Stern–Gerlach experiment demonstrated that the spatial orientation of angular momentum is quantized. Thus an atomic-scale system was shown to have intrinsically quantum properties. In the original experiment, silver atoms were sent through a spatially varying magnetic field, which deflected them before they struck a detector screen, such as a glass slide. Particles with non-zero magnetic moment are deflected, due to the magnetic field gradient, from a straight path. The screen reveals discrete points of accumulation, rather than a continuous distribution, owing to their quantized spin. Historically, this experiment was decisive in convincing physicists of the reality of angular-momentum quantization in all atomic-scale systems. The experiment was first conducted by the German physicists Otto Stern and Walter Gerlach in 1922.

The Stern–Gerlach experiment involves sending a beam of silver atoms through an inhomogeneous magnetic field and observing their deflection. The results show that particles possess an intrinsic angular momentum that is closely analogous to the angular momentum of a classically spinning object, but that takes only certain quantized values. Another important result is that only one component of a particle's spin can be measured at one time, meaning that the measurement of the spin along the z-axis destroys information about a particle's spin along the x and y axis.

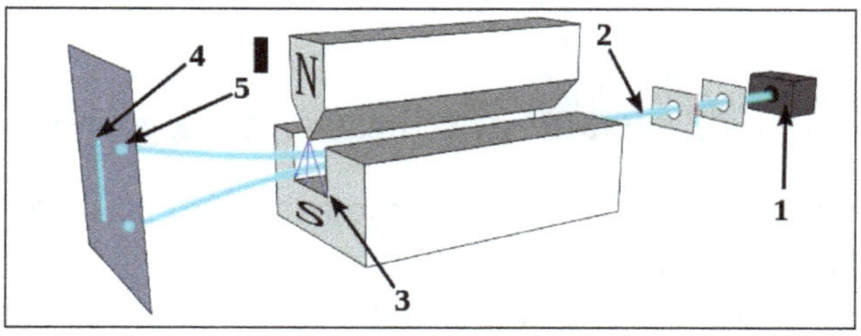

Stern–Gerlach experiment: Silver atoms travelling through an inhomogeneous magnetic field, and being deflected up or down depending on their spin; (1) furnace, (2) beam of silver atoms, (3) inhomogeneous magnetic field, (4) classically expected result, (5) observed result.

The experiment is normally conducted using electrically neutral particles such as silver atoms. This avoids the large deflection in the path of a charged particle moving through a magnetic field and allows spin-dependent effects to dominate.

If the particle is treated as a classical spinning magnetic dipole, it will precess in a magnetic field because of the torque that the magnetic field exerts on the dipole. If it moves through a homogeneous magnetic field, the forces exerted on opposite ends of the dipole cancel each other out and the trajectory of the particle is unaffected. However, if the magnetic field is inhomogeneous then the force on one end of the dipole will be slightly greater than the opposing force on the other end, so that there is a net force which deflects the particle's trajectory. If the particles were classical spinning objects, one would expect the distribution of their spin angular momentum vectors to be random and continuous. Each particle would be deflected by an amount proportional to its magnetic moment, producing some density distribution on the detector screen. Instead, the particles passing through the Stern–Gerlach apparatus are deflected either up or down by a specific amount. This was a measurement of the quantum observable now known as spin angular momentum, which demonstrated possible outcomes of a measurement where the observable has a discrete set of values or point spectrum.

Although some discrete quantum phenomena, such as atomic spectra, were observed much earlier, the Stern–Gerlach experiment allowed scientists to observe separation between discrete quantum states for the first time in the history of science. By now, it is known that, theoretically, quantum angular momentum of any kind has a discrete spectrum, which is sometimes briefly expressed as "angular momentum is quantized".

Experiment using Particles with + ½ or – ½ Spin

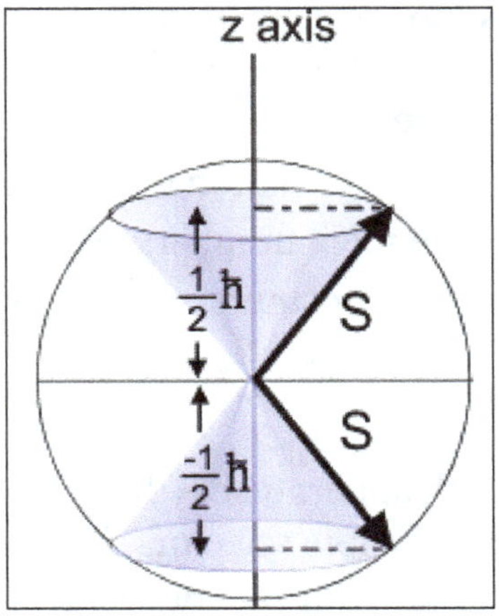

Spin values for fermions.

If the experiment is conducted using charged particles like electrons, there will be a Lorentz force that tends to bend the trajectory in a circle. This force can be cancelled by an electric field of appropriate magnitude oriented transverse to the charged particle's path.

Electrons are spin-½ particles. These have only two possible spin angular momentum values measured along any axis, $+\frac{\hbar}{2}$ or $-\frac{\hbar}{2}$, a purely quantum mechanical phenomenon. Because its value is always the same, it is regarded as an intrinsic property of electrons, and is sometimes known as "intrinsic angular momentum" (to distinguish it from orbital angular momentum, which can vary and depends on the presence of other particles). If one measures the spin along a vertical axis, electrons are described as "spin up" or "spin down", based on the magnetic moment pointing up or down, respectively.

To mathematically describe the experiment with spin $+\frac{1}{2}$ particles, it is easiest to use Dirac's bra–ket notation. As the particles pass through the Stern–Gerlach device, they are deflected either up or down, and observed by the detector which resolves to either spin up or spin down. These are described by the angular momentum quantum number j, which can take on one of the two possible allowed values, either $+\frac{\hbar}{2}$ or $-\frac{\hbar}{2}$. The act of observing (measuring) the momentum along the z axis corresponds to the operator j_z. [specify] In mathematical terms, the initial state of the particles is:

$$|\psi\rangle = c_1 \left|\psi_{j=+\frac{\hbar}{2}}\right\rangle + c_2 \left|\psi_{j=-\frac{\hbar}{2}}\right\rangle$$

where constants c_1 and c_2 are complex numbers. This initial state spin can point in any direction. The squares of the absolute values $|c_1|^2$ and $|c_2|^2$ determine the probabilities that for a system in the initial state $|\psi\rangle$ one of the two possible values of j is found after the measurement is made. The constants c_1 and c_2 must also be normalized in order that the probability of finding either one of the values be unity, that is we must ensure that $|c_1|^2 + |c_2|^2 = 1$. However, this information is not sufficient to determine the values of c_1 and c_2, because they are complex numbers. Therefore, the measurement yields only the squared magnitudes of the constants, which are interpreted as probabilities.

Sequential Experiments

If we link multiple Stern–Gerlach apparatuses (the rectangles containing S-G), we can clearly see that they do not act as simple selectors, i.e. filtering out (blocking) particles with one of the states and leaving the others unchanged. Instead they alter the states observed (as in light polarization). In the figure below x and z name the direction of the (inhomogenous) magnetic field, with the x-z-plane being orthogonal to the particle beam. In the three S-G systems shown below, the cross-hatched squares denote the blocking of a given output, i.e. each of the S-G systems with a blocker allows only particles with one of two states to enter the next S-G apparatus in the sequence.

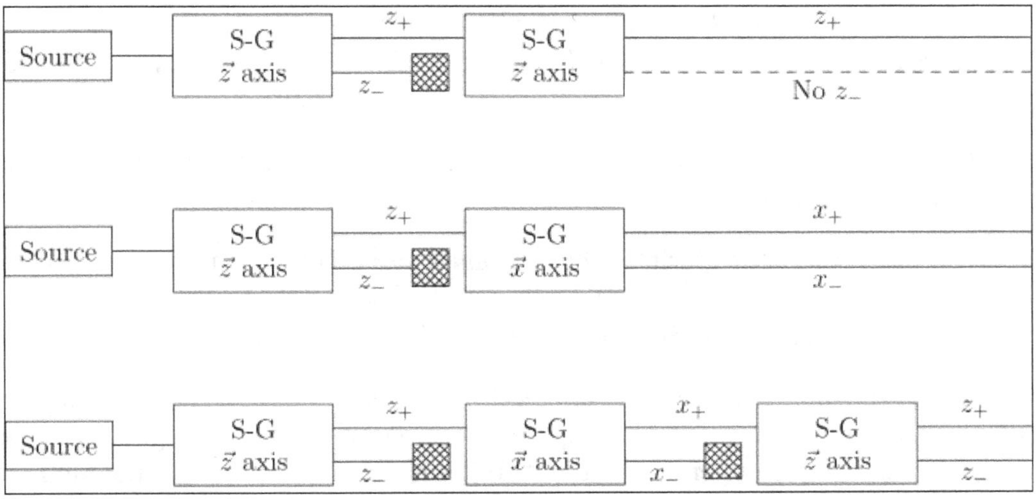

The top system demonstrates that when z+ is passed through a second, identical, S-G apparatus, only z+ is seen in the output, as expected. The middle system demonstrates that a z+ input can be sorted on the x-axis producing x+ and x- outputs. The bottom system feeds the x+ output into a third S-G apparatus and demonstrates that it be sorted on the z-axis to produce z+ and z- outputs. Given that the input to the second S-G apparatus consisted only of z+, while the output of the third contains z- as well as z+, it can be inferred that an S-G apparatus must be altering the states of the particles that pass through it.

Importance

The Stern–Gerlach experiment strongly influenced later developments in modern physics:

- In the decade that followed, scientists showed using similar techniques, that the nuclei of some atoms also have quantized angular momentum. It is the interaction of this nuclear angular momentum with the spin of the electron that is responsible for the hyperfine structure of the spectroscopic lines.
- In the 1930s, using an extended version of the Stern–Gerlach apparatus, Isidor Rabi and colleagues showed that by using a varying magnetic field, one can force the magnetic moment to go from one state to the other. The series of experiments culminated in 1937 when they discovered that state transitions could be induced using time varying fields or RF fields. The so-called Rabi oscillation is the working mechanism for the Magnetic Resonance Imaging equipment found in hospitals.

- Norman F. Ramsey later modified the Rabi apparatus to increase the interaction time with the field. The extreme sensitivity due to the frequency of the radiation makes this very useful for keeping accurate time, and it is still used today in atomic clocks.
- In the early sixties, Ramsey and Daniel Kleppner used a Stern–Gerlach system to produce a beam of polarized hydrogen as the source of energy for the hydrogen Maser, which is still one of the most popular atomic clocks.
- The direct observation of the spin is the most direct evidence of quantization in quantum mechanics.
- The Stern–Gerlach experiment has become a paradigm of quantum measurement.

Quantum Mechanical Interpretation of Stern-Gerlach Experiment

For an atom, there are three possible magnetic moment source: magnetic moment due to electrons' orbital angular momenta, spin of electrons and nuclear magnetic moment. The magnitude of these depends inversely on the mass of the source of magnetic moment, $\mu \propto \dfrac{e\hbar}{2m}$. So electronic magnetic moments and nuclear magnetic moments differ in order, and nuclear magnetic moment is much smaller than electronic magnetic moments due to nucleonelectron mass ratio $m_N/m_e \approx 2000$. This ratio tells us that the deflection in the Stern-Gerlach experiment is originating from electron.

There are two reason for magnetic moment arising from electron; orbital angular momenta and spin. Before looking through to these, we should look through electron's description. Firstly, the electron is considered as a point particle. In quantum mechanics without any causal reason, except consistency with the experiments, we use four quantum numbers to define an electron orbiting around the nucleus: principal quantum number n, orbital angular momentum quantum number l, magnetic quantum number m_l and electron spin quantum number m_s.

One of the electronic magnetic moment is related with orbital motion of electrons and it can be calculated by using m_l. It takes integer values between –l and l, in total 2l + 1 values. Since l is always integer, there must be odd number of possible m_l values. Other electronic source of atomic magnetic moment is electron spin. In quantum mechanics, electron spin is half integer and there are only two possible corresponding m_s values, $-1/2$ and $1/2$. If an atom has an uncoupled electron at the outer shell, this electron cancause magnetic moment, other coupled ones in full shells do not contribute to atom's magnetic moment since they cancel each other's effect. If we want to calculate magnetic moment of an atom due to its electrons, we should only consider unfilled shells. The electrons' contribution to the magnetic moment of the atom comes from electrons in the unfilled outer shells and can be calculated from their m_l and m_s values. This contribution is two fold; due to orbital motion and electron spin. This is the reason why in SternGerlach experiments with different atoms one can obtain different patterns. We already explained silver atom case with $m_l = 0$ and $m_s = \pm 1/2$. Since m_l is zero, it does not contribute to magnetic moment and there are two possible m_s value each corresponding one of the traces.

Half integer value of the electron spin is firstly invented to be able to obtain a framework which is consistent with the experiments. There were two different traces, and electron spin is considered as 1/2 to be able to get consistent structure with quantum mechanics, in which the increment of quantum numbers is always 1. The remaining multiplication factors, multiplication of g-factor for electron spin and μ_b, are arranged to be able to obtain consistent results with experiment.

References

- *Assadi, M.H.N; Hanaor, D.A.H (2013).* "Theoretical study on copper's energetics and magnetism in TiO2 polymorphs". *Journal of Applied Physics. 113 (23): 233913–233913–5.* arXiv:1304.1854. Bibcode:*2013JAP...113w3913A*. doi:10.1063/1.4811539
- Angular-momentum, apeer: physics.ucc.ie, Retrieved 10 July, 2020
- "CODATA Value: electron g factor". *The NIST Reference on Constants, Units, and Uncertainty*. NIST. *2018*. Retrieved 2019-06-04
- A., Henok (2002). Introduction to Applied Modern Physics. Lulu.com. p. 76. ISBN 1-4357-0521-1

6

Theories and Methods

Approximation methods are useful for approximations to wave functions, energies, spectra, as well as transition probabilities and other dynamical quantities. Scattering theory deals with the study of the scattering of waves and particles. This chapter sheds light on the methods and theories used in quantum mechanics for an in-depth understanding of the subject.

The exact solution of Schrödinger equation can be found for only a limited number of problems, such as square wells, harmonic oscillator, hydrogen atom etc. A majority of problems in quantum mechanics can usually be solved using a wide variety of approximate methods.

Perturbation theory applies to the cases where the Hamiltonian is only slightly different from the Hamiltonian that can be exactly solved. When this difference or perturbation is small then the approximate wavefunction and energy spectrum are obtained by adding appropriate corrections to the wavefunction and energy of the exactly solvable Hamiltonian. But often the Hamiltonian cannot be reduced to an exactly solvable and perturbation parts and in those cases variational method becomes useful. It gives us fairly good estimates of the energies of the ground state and a first few excited states for which one has only qualitative knowledge of the wavefunctions.

Approximation methods have a practical and a conceptual value. On the practical side, because all but the simplest systems resist attempts to find closed form analytical solutions to eigenvalue problems, the Schrödinger equation, etc., we use such methods to get useful, approximations to wave functions, energies, spectra, as well as transition probabilities and other dynamical quantities. On the conceptual side we shall see that some of our most cherished ways of thinking about energy levels and dynamics stem principally from the point of view of approximation methods.

The need for approximation methods arises from the simple fact that almost all realistic physical systems one wants to study are too complicated for explicit analytic solutions to be available. So, for example, while we can analytically handle the hydrogen atom (when modeled as a charged particle in a Coulomb field), we cannot handle helium or more complicated atoms in the same fashion – let alone dynamical processes involving the interaction of these atoms with electromagnetic fields. In fact, even more realistic models of the hydrogen atom (including things like spin-orbit coupling, hyperfine interaction, finite size of nucleus, etc.) are not exactly soluble. Thus the only way we can understand these systems is to find methods of approximation.

PERTURBATION THEORY

In quantum mechanics, perturbation theory is a set of approximation schemes directly related to mathematical perturbation for describing a complicated quantum system in terms of a simpler one. The idea is to start with a simple system for which a mathematical solution is known, and add an additional "perturbing" Hamiltonian representing a weak disturbance to the system. If the disturbance is not too large, the various physical quantities associated with the perturbed system (e.g. its energy levels and eigenstates) can be expressed as "corrections" to those of the simple system. These corrections, being small compared to the size of the quantities themselves, can be calculated using approximate methods such as asymptotic series. The complicated system can therefore be studied based on knowledge of the simpler one. In effect, it is describing a complicated unsolved system using a simple, solved system.

Approximate Hamiltonians

Perturbation theory is an important tool for describing real quantum systems, as it turns out to be very difficult to find exact solutions to the Schrödinger equation for Hamiltonians of even moderate complexity. The Hamiltonians to which we know exact solutions, such as the hydrogen atom, the quantum harmonic oscillator and the particle in a box, are too idealized to adequately describe most systems. Using perturbation theory, we can use the known solutions of these simple Hamiltonians to generate solutions for a range of more complicated systems.

Applying Perturbation Theory

Perturbation theory is applicable if the problem at hand cannot be solved exactly, but can be formulated by adding a "small" term to the mathematical description of the exactly solvable problem. For example, by adding a perturbative electric potential to the quantum mechanical model of the hydrogen atom, tiny shifts in the spectral lines of hydrogen caused by the presence of an electric field (the Stark effect) can be calculated. This is only approximate because the sum of a Coulomb potential with a linear potential is unstable (has no true bound states) although the tunneling time (decay rate) is very long. This instability shows up as a broadening of the energy spectrum lines, which perturbation theory fails to reproduce entirely.

The expressions produced by perturbation theory are not exact, but they can lead to accurate results as long as the expansion parameter, say α, is very small. Typically, the results are expressed in terms of finite power series in α that seem to converge to the exact values when summed to higher order. After a certain order $n \sim 1/\alpha$ however, the results become increasingly worse since the series are usually divergent (being asymptotic series). There exist ways to convert them into convergent series, which can be evaluated for large-expansion parameters, most efficiently by the variational method.

In the theory of quantum electrodynamics (QED), in which the electron–photon interaction is treated perturbatively, the calculation of the electron's magnetic moment has been found to agree with experiment to eleven decimal places. In QED and other quantum field theories, special calculation techniques known as Feynman diagrams are used to systematically sum the power series terms.

Limitations

Large Perturbations

Under some circumstances, perturbation theory is an invalid approach to take. This happens when the system we wish to describe cannot be described by a small perturbation imposed on some simple system. In quantum chromodynamics, for instance, the interaction of quarks with the gluon field cannot be treated perturbatively at low energies because the coupling constant (the expansion parameter) becomes too large.

Non-adiabatic States

Perturbation theory also fails to describe states that are not generated adiabatically from the "free model", including bound states and various collective phenomena such as solitons. Imagine, for example, that we have a system of free (i.e. non-interacting) particles, to which an attractive interaction is introduced. Depending on the form of the interaction, this may create an entirely new set of eigenstates corresponding to groups of particles bound to one another. An example of this phenomenon may be found in conventional superconductivity, in which the phonon-mediated attraction between conduction electrons leads to the formation of correlated electron pairs known as Cooper pairs. When faced with such systems, one usually turns to other approximation schemes, such as the variational method and the WKB approximation. This is because there is no analogue of a bound particle in the unperturbed model and the energy of a soliton typically goes as the *inverse* of the expansion parameter. However, if we "integrate" over the solitonic phenomena, the nonperturbative corrections in this case will be tiny; of the order of $\exp(-1/g)$ or $\exp(-1/g^2)$ in the perturbation parameter g. Perturbation theory can only detect solutions "close" to the unperturbed solution, even if there are other solutions for which the perturbative expansion is not valid.

Difficult Computations

The problem of non-perturbative systems has been somewhat alleviated by the advent of modern computers. It has become practical to obtain numerical non-perturbative solutions for certain problems, using methods such as density functional theory. These advances have been of particular benefit to the field of quantum chemistry. Computers have also been used to carry out perturbation theory calculations to extraordinarily high levels of precision, which has proven important in particle physics for generating theoretical results that can be compared with experiment.

Time-independent Perturbation Theory

Time-independent perturbation theory is one of two categories of perturbation theory, the other being time-dependent perturbation. In time-independent perturbation theory the perturbation Hamiltonian is static (i.e., possesses no time dependence). Time-independent perturbation theory was presented by Erwin Schrödinger in a 1926 paper, shortly after he produced his theories in wave mechanics. In this paper Schrödinger referred to earlier work of Lord Rayleigh, who investigated harmonic vibrations of a string perturbed by small inhomogeneities. This is why this perturbation theory is often referred to as Rayleigh–Schrödinger perturbation theory.

First Order Corrections

We begin with an unperturbed Hamiltonian H_0, which is also assumed to have no time dependence. It has known energy levels and eigenstates, arising from the time-independent Schrödinger equation:

$$H_0 \left| n^{(0)} \right\rangle = E_n^{(0)} \left| n^{(0)} \right\rangle, \qquad n = 1,2,3,\cdots$$

For simplicity, we have assumed that the energies are discrete. The (0) superscripts denote that these quantities are associated with the unperturbed system. Note the use of bra–ket notation.

We now introduce a perturbation to the Hamiltonian. Let V be a Hamiltonian representing a weak physical disturbance, such as a potential energy produced by an external field. (Thus, V is formally a Hermitian operator). Let λ be a dimensionless parameter that can take on values ranging continuously from 0 (no perturbation) to 1 (the full perturbation). The perturbed Hamiltonian is:

$$H = H_0 + \lambda V$$

The energy levels and eigenstates of the perturbed Hamiltonian are again given by the Schrödinger equation,

$$\left(H_0 + \lambda V\right) | n\rangle = E_n | n\rangle.$$

Our goal is to express E_n and $|n\rangle$ in terms of the energy levels and eigenstates of the old Hamiltonian. If the perturbation is sufficiently weak, we can write them as a (Maclaurin) power series in λ,

$$E_n = E_n^{(0)} + \lambda E_n^{(1)} + \lambda^2 E_n^{(2)} + \cdots$$
$$|n\rangle = |n^{(0)}\rangle + \lambda |n^{(1)}\rangle + \lambda^2 |n^{(2)}\rangle + \cdots$$

where

$$E_n^{(k)} = \frac{1}{k!} \frac{d^k E_n}{d\lambda^k} \Big|_{\lambda=0}$$
$$|n^{(k)}\rangle = \frac{1}{k!} \frac{d^k |n\rangle}{d\lambda^k} \Big|_{\lambda=0.}$$

When $k = 0$, these reduce to the unperturbed values, which are the first term in each series. Since the perturbation is weak, the energy levels and eigenstates should not deviate too much from their unperturbed values, and the terms should rapidly become smaller as we go to higher order.

Substituting the power series expansion into the Schrödinger equation, we obtain:

$$(H_0 + \lambda V)\left(|n^{(0)}\rangle + \lambda |n^{(1)}\rangle + \cdots\right) = \left(E_n^{(0)} + \lambda E_n^{(1)} + \cdots\right)\left(|n^{(0)}\rangle + \lambda |n^{(1)}\rangle + \cdots\right).$$

Expanding this equation and comparing coefficients of each power of λ results in an infinite series of simultaneous equations. The zeroth-order equation is simply the Schrödinger equation for the unperturbed system.

The first-order equation is:

$$H_0 |n^{(1)}\rangle + V |n^{(0)}\rangle = E_n^{(0)} |n^{(1)}\rangle + E_n^{(1)} |n^{(0)}\rangle.$$

Operating through by $\langle n^{(0)}|$, the first term on the left-hand side cancels the first term on the right-hand side. (Recall, the unperturbed Hamiltonian is Hermitian). This leads to the first-order energy shift,

$$E_n^{(1)} = \langle n^{(0)} | V | n^{(0)} \rangle.$$

This is simply the expectation value of the perturbation Hamiltonian while the system is in the unperturbed state.

This result can be interpreted in the following way: suppose the perturbation is applied, but we keep the system in the quantum state $|n^{(0)}\rangle$, which is a valid quantum state though no longer an energy eigenstate. The perturbation causes the average energy of this state to increase by $\langle n^{(0)} | V | n^{(0)}\rangle$. However, the true energy shift is slightly different, because the perturbed eigenstate is not exactly the same as $|n^{(0)}\rangle$. These further shifts are given by the second and higher order corrections to the energy.

Before we compute the corrections to the energy eigenstate, we need to address the issue of normalization. We may suppose:

$$\left\langle n^{(0)} \mid n^{(0)} \right\rangle = 1,$$

but perturbation theory assumes we also have $\langle n \mid n \rangle = 1$.

It follows that at first order in λ, we must have:

$$\left(\langle n^{(0)} \mid + \lambda \langle n^{(1)} \mid\right)\left(\left| n^{(0)} \right\rangle + \lambda \left| n^{(1)} \right\rangle\right) = 1$$

$$\left\langle n^{(0)} \mid n^{(0)} \right\rangle + \lambda \left\langle n^{(0)} \mid n^{(1)} \right\rangle \mid + \lambda \left\langle n^{(1)} \mid n^{(0)} \right\rangle + \cancel{\lambda^2 \left\langle n^{(1)} \mid n^{(1)} \right\rangle} = 1$$

$$\left\langle n^{(0)} \mid n^{(1)} \right\rangle + \left\langle n^{(1)} \mid n^{(0)} \right\rangle = 0.$$

Since the overall phase is not determined in quantum mechanics, without loss of generality, in time independent theory we may assume $\langle n^{(0)} \mid n^{(1)} \rangle$, is purely real. Therefore,

$$\left\langle n^{(0)} \mid n^{(1)} \right\rangle = \left\langle n^{(1)} \mid n^{(0)} \right\rangle$$

leading to,

$$\left\langle n^{(0)} \mid n^{(1)} \right\rangle = 0.$$

To obtain the first-order correction to the energy eigenstate, we insert our expression for the first-order energy correction back into the result shown above equating the first-order coefficients of λ. We then make use of the resolution of the identity,

$$\begin{aligned} V\left| n^{(0)} \right\rangle &= \left(\sum_{k \neq n} \left| k^{(0)} \right\rangle \langle k^{(0)} \mid \right) V \left| n^{(0)} \right\rangle + \left(\left| n^{(0)} \right\rangle \langle n^{(0)} \mid \right) V \left| n^{(0)} \right\rangle \\ &= \sum_{k \neq n} \left| k^{(0)} \right\rangle \langle k^{(0)} \mid V \left| n^{(0)} \right\rangle + E_n^{(1)} \left| n^{(0)} \right\rangle, \end{aligned}$$

where the $\mid k^{(0)} \rangle$ are in the orthogonal complement of $\mid n^{(0)} \rangle$.

The first-order equation may thus be expressed as:

$$\left(E_n^{(0)} - H_0\right)\left| n^{(1)} \right\rangle = \sum_{k \neq n} \left| k^{(0)} \right\rangle \left\langle k^{(0)} \middle| V \middle| n^{(0)} \right\rangle.$$

For the moment, suppose that the zeroth-order energy level is not degenerate, i.e. there is no eigenstate of H_0 in the orthogonal complement of $\mid n^{(0)} \rangle$ with the energy $E_n^{(0)}$. After renaming the summation dummy index above as k', we can pick any $k \neq n$, and multiply through by $\langle k^{(0)} \mid$ giving:

$$\left(E_n^{(0)} - E_k^{(0)}\right) k^{(0)} \rangle \left| n^{(1)} \right\rangle = \left\langle k^{(0)} \middle| V \middle| n^{(0)} \right\rangle.$$

We see that the above $\langle k^{(0)} \mid n^{(1)}\rangle$ also gives us the component of the first-order correction along $\mid k^{(0)}\rangle$.

Thus, in total, we get,

$$\left|n^{(1)}\right\rangle = \sum_{k\neq n} \frac{\langle k^{(0)} \mid V \left| n^{(0)} \right\rangle}{E_n^{(0)} - E_k^{(0)}} \left| k^{(0)} \right\rangle.$$

The first-order change in the n-th energy eigenket has a contribution from each of the energy eigenstates $k \neq n$. Each term is proportional to the matrix element $\langle k^{(0)} \mid V \mid n^{(0)}\rangle$, which is a measure of how much the perturbation mixes eigenstate n with eigenstate k; it is also inversely proportional to the energy difference between eigenstates k and n, which means that the perturbation deforms the eigenstate to a greater extent if there are more eigenstates at nearby energies. The expression is singular if any of these states have the same energy as state n, which is why we assumed that there is no degeneracy.

Second-order and Higher Corrections

We can find the higher-order deviations by a similar procedure, though the calculations become quite tedious with our current formulation. Our normalization prescription gives that:

$$2\langle n^{(0)} \mid \left\langle n^{(2)} + \langle n^{(1)} \mid \right\langle n^{(1)} = 0.$$

Up to second order, the expressions for the energies and (normalized) eigenstates are:

$$E_n(\lambda) = E_n^{(0)} + \lambda \langle n^{(0)} \mid V \left| n^{(0)} \right\rangle + \lambda^2 \sum_{k\neq n} \frac{\left|\langle k^{(0)} \mid V \left| n^{(0)} \right\rangle\right|^2}{E_n^{(0)} - E_k^{(0)}} + O(\lambda^3)$$

$$\mid n(\lambda)\rangle = \left|n^{(0)}\right\rangle + \lambda \sum_{k\neq n} \left|k^{(0)}\right\rangle \frac{\langle k^{(0)} \mid V \left| n^{(0)}\right\rangle}{E_n^{(0)} - E_k^{(0)}} + \lambda^2 \sum_{k\neq n}\sum_{\ell\neq n} \left|k^{(0)}\right\rangle \frac{\langle k^{(0)} \mid V \left|\ell^{(0)}\right\rangle \langle \ell^{(0)} \mid V \left|n^{(0)}\right\rangle}{\left(E_n^{(0)} - E_k^{(0)}\right)\left(E_n^{(0)} - E_\ell^{(0)}\right)}$$

$$-\lambda^2 \sum_{k\neq n} \left|k^{(0)}\right\rangle \frac{\langle n^{(0)} \mid V \left|n^{(0)}\right\rangle \langle k^{(0)} \mid V \left|n^{(0)}\right\rangle}{\left(E_n^{(0)} - E_k^{(0)}\right)^2} - \frac{1}{2}\lambda^2 \left|n^{(0)}\right\rangle \sum_{k\neq n} \frac{\langle n^{(0)} \mid V \left|k^{(0)}\right\rangle \langle k^{(0)} \mid V \left|n^{(0)}\right\rangle}{\left(E_n^{(0)} - E_k^{(0)}\right)^2} + O(\lambda^3).$$

Extending the process further, the third-order energy correction can be shown to be:

$$E_n^{(3)} = \sum_{k\neq n}\sum_{m\neq n} \frac{\langle n^{(0)} \mid V \mid m^{(0)}\rangle \langle m^{(0)} \mid V \mid k^{(0)}\rangle \langle k^{(0)} \mid V \mid n^{(0)}\rangle}{\left(E_n^{(0)} - E_m^{(0)}\right)\left(E_n^{(0)} - E_k^{(0)}\right)} - \langle n^{(0)} \mid V \mid n^{(0)}\rangle \sum_{m\neq n} \frac{\mid \langle n^{(0)} \mid V \mid m^{(0)}\rangle \mid^2}{\left(E_n^{(0)} - E_m^{(0)}\right)^2}.$$

Effects of Degeneracy

Suppose that two or more energy eigenstates are degenerate. The first-order energy shift is not well defined, since there is no unique way to choose a basis of eigenstates for the unperturbed system. The various eigenstates for a given energy will perturb with different energies, or may well possess no continuous family of perturbations at all.

This is manifested in the calculation of the perturbed eigenstate via the fact that the operator:

$$E_n^{(0)} - H_0$$

does not have a well-defined inverse.

Let D denote the subspace spanned by these degenerate eigenstates. No matter how small the perturbation is, in the degenerate subspace D the energy differences between the eigenstates of H are non-zero, so complete mixing of at least some of these states is assured. Typically, the eigenvalues will split, and the eigenspaces will become simple (one-dimensional), or at least of smaller dimension than D.

The successful perturbations will not be "small" relative to a poorly chosen basis of D. Instead, we consider the perturbation "small" if the new eigenstate is close to the subspace D. The new Hamiltonian must be diagonalized in D, or a slight variation of D, so to speak. These perturbed eigenstates in D are now the basis for the perturbation expansion,

$$|n\rangle = \sum_{k \in D} \alpha_{nk} |k^{(0)}\rangle + \lambda |n^{(1)}\rangle.$$

For the first-order perturbation, we need solve the perturbed Hamiltonian restricted to the degenerate subspace D,

$$V|k^{(0)}\rangle = \epsilon_k |k^{(0)}\rangle + \text{small} \qquad \forall |k^{(0)}\rangle \in D,$$

simultaneously for all the degenerate eigenstates, where ϵ_k are first-order corrections to the degenerate energy levels, and "small" is a vector of $O(\lambda)$ orthogonal to D. This amounts to diagonalizing the matrix:

$$\langle k^{(0)} | V | l^{(0)} \rangle = V_{kl} \qquad \forall |k^{(0)}\rangle, |l^{(0)}\rangle \in D.$$

This procedure is approximate, since we neglected states outside the D subspace ("small"). The splitting of degenerate energies ϵ_k is generally observed. Although the splitting may be small, $O(\lambda)$, compared to the range of energies found in the system, it is crucial in understanding certain details, such as spectral lines in Electron Spin Resonance experiments.

Higher-order corrections due to other eigenstates outside D can be found in the same way as for the non-degenerate case,

$$\left(E_n^{(0)} - H_0\right)|n^{(1)}\rangle = \sum_{k \notin D} \left(\langle k^{(0)} | V | n^{(0)} \rangle\right) |k^{(0)}\rangle$$

The operator on the left-hand side is not singular when applied to eigenstates outside D, so we can write:

$$|n^{(1)}\rangle = \sum_{k\notin D} \frac{\langle k^{(0)}|V|n^{(0)}\rangle}{E_n^{(0)} - E_k^{(0)}} |k^{(0)}\rangle,$$

but the effect on the degenerate states is of $O(\lambda)$.

Near-degenerate states should also be treated similarly, when the original Hamiltonian splits aren't larger than the perturbation in the near-degenerate subspace. An application is found in the nearly free electron model, where near-degeneracy, treated properly, gives rise to an energy gap even for small perturbations. Other eigenstates will only shift the absolute energy of all near-degenerate states simultaneously.

Generalization to Multi-parameter Case

The generalization of the time-independent perturbation theory to the case where there are multiple small parameters $x^\mu = (x^1, x^2, \cdots)$ in place of λ can be formulated more systematically using the language of differential geometry, which basically defines the derivatives of the quantum states and calculates the perturbative corrections by taking derivatives iteratively at the unperturbed point.

Hamiltonian and Force Operator

From the differential geometric point of view, a parameterized Hamiltonian is considered as a function defined on the parameter manifold that maps each particular set of parameters $(x^1, x^2, \cdots)$ to an Hermitian operator $H(x^\mu)$ that acts on the Hilbert space. The parameters here can be external field, interaction strength, or driving parameters in the quantum phase transition. Let $E_n(x^\mu)$ and $|n(x^\mu)\rangle$ be the n-th eigenenergy and eigenstate of $H(x^\mu)$ respectively. In the language of differential geometry, the states $|n(x^\mu)\rangle$ form a vector bundle over the parameter manifold, on which derivatives of these states can be defined. The perturbation theory is to answer the following question: given $E_n(x_0^\mu)$ and $|n(x_0^\mu)\rangle$ at an unperturbed reference point x_0^μ, how to estimate the $E_n(x^\mu)$ and $|n(x^\mu)\rangle$ at x^μ close to that reference point.

Without loss of generality, the coordinate system can be shifted, such that the reference point $x_0^\mu = 0$ is set to be the origin. The following linearly parameterized Hamiltonian is frequently used,

$$H(x^\mu) = H(0) + x^\mu F_\mu.$$

If the parameters x^μ are considered as generalized coordinates, then F_μ should be identified as the generalized force operators related to those coordinates. Different indices μ label the different forces along different directions in the parameter manifold. For example, if x^μ denotes the external magnetic field in the μ-direction, then F_μ should be the magnetization in the same direction.

Perturbation Theory as Power Series Expansion

The validity of the perturbation theory lies on the adiabatic assumption, which assumes the eigenenergies and eigenstates of the Hamiltonian are smooth functions of parameters such that their values in the vicinity region can be calculated in power series (like Taylor expansion) of the parameters:

$$E_n(x^\mu) = E_n + x^\mu \partial_\mu E_n + \frac{1}{2!} x^\mu x^\nu \partial_\mu \partial_\nu E_n + \cdots$$

$$\left| n(x^\mu) \right\rangle = | n\rangle + x^\mu \, | \partial_\mu n\rangle + \frac{1}{2!} x^\mu x^\nu \, | \partial_\mu \partial_\nu n\rangle + \cdots$$

Here ∂_μ denotes the derivative with respect to x^μ. When applying to the state $| \partial_\mu n\rangle$, it should be understood as the covariant derivative if the vector bundle is equipped with non-vanishing connection. All the terms on the right-hand-side of the series are evaluated at $x^\mu = 0$, e.g. $E_n \equiv E_n(0)$ and $| n\rangle \equiv | n(0)\rangle$. This convention will be adopted throughout this subsection, that all functions without the parameter dependence explicitly stated are assumed to be evaluated at the origin. The power series may converge slowly or even not converge when the energy levels are close to each other. The adiabatic assumption breaks down when there is energy level degeneracy, and hence the perturbation theory is not applicable in that case.

Hellmann–feynman Theorems

The above power series expansion can be readily evaluated if there is a systematic approach to calculate the derivates to any order. Using the chain rule, the derivatives can be broken down to the single derivative on either the energy or the state. The Hellmann–Feynman theorems are used to calculate these single derivatives. The first Hellmann–Feynman theorem gives the derivative of the energy,

$$\partial_\mu E_n = \langle n | \partial_\mu H | n\rangle$$

The second Hellmann–Feynman theorem gives the derivative of the state (resolved by the complete basis with m ≠ n),

$$\langle m | \partial_\mu n\rangle = \frac{\langle m | \partial_\mu H | n\rangle}{E_n - E_m}, \qquad \langle \partial_\mu m | n\rangle = \frac{\langle m | \partial_\mu H | n\rangle}{E_m - E_n}.$$

For the linearly parameterized Hamiltonian, $\partial_\mu H$ simply stands for the generalized force operator F_μ.

The theorems can be simply derived by applying the differential operator ∂_μ to both sides of the Schrödinger equation $H | n\rangle = E_n | n\rangle$, which reads:

$$\partial_\mu H | n\rangle + H | \partial_\mu n\rangle = \partial_\mu E_n | n\rangle + E_n | \partial_\mu n\rangle \cdot$$

Then overlap with the state $\langle m|$ from left and make use of the Schrödinger equation $\langle m|H = \langle m|E_m$ again,

$$\langle m|\partial_\mu H|n\rangle + E_m\langle m|\partial_\mu n\rangle = \partial_\mu E_n\langle m|n\rangle + E_n\langle m|\partial_\mu n\rangle$$

Given that the eigenstates of the Hamiltonian always form an orthonormal basis $\langle m|n\rangle = \delta_{mn}$, the cases of $m = n$ and $m \neq n$ can be discussed separately. The first case will lead to the first theorem and the second case to the second theorem, which can be shown immediately by rearranging the terms. With the differential rules given by the Hellmann–Feynman theorems, the perturbative correction to the energies and states can be calculated systematically.

Correction of Energy and State

To the second order, the energy correction reads:

$$E_n(x^\mu) = \langle n|H|n\rangle + \langle n|\partial_\mu H|n\rangle x^\mu + \Re\sum_{m\neq n}\frac{\langle n|\partial_\nu H|m\rangle\langle m|\partial_\mu H|n\rangle}{E_n - E_m}x^\mu x^\nu + \cdots,$$

where $\Re$ denotes the real part function. The first order derivative $\partial_\mu E_n$ is given by the first Hellmann–Feynman theorem directly. To obtain the second order derivative $\partial_\mu\partial_\nu E_n$, simply applying the differential operator ∂_μ to the result of the first order derivative $\langle n|\partial_\nu H|n\rangle$, which reads:

$$\partial_\mu\partial_\nu E_n = \langle\partial_\mu n|\partial_\nu H|n\rangle + \langle n|\partial_\mu\partial_\nu H|n\rangle + \langle n|\partial_\nu H|\partial_\mu n\rangle$$

Note that for linearly parameterized Hamiltonian, there is no second derivative $\partial_\mu\partial_\nu H = 0$ on the operator level. Resolve the derivative of state by inserting the complete set of basis,

$$\partial_\mu\partial_\nu E_n = \sum_m\big(\langle\partial_\mu n|m\rangle\langle m|\partial_\nu H|n\rangle + \langle n|\partial_\nu H|m\rangle\langle m|\partial_\mu n\rangle\big),$$

then all parts can be calculated using the Hellmann–Feynman theorems. In terms of Lie derivatives, $\langle\partial_\mu n|n\rangle = \langle n|\partial_\mu n\rangle = 0$ according to the definition of the connection for the vector bundle. Therefore, the case $m = n$ can be excluded from the summation, which avoids the singularity of the energy denominator. The same procedure can be carried on for higher order derivatives, from which higher order corrections are obtained.

The same computational scheme is applicable for the correction of states. The result to the second order is as follows:

$$\begin{aligned}|n(x^\mu)\rangle =& |n\rangle + \sum_{m\neq n}\frac{\langle m|\partial_\mu H|n\rangle}{E_n - E_m}|m\rangle x^\mu \\ &+\left(\sum_{m\neq n}\sum_{l\neq n}\frac{\langle m|\partial_\mu H|l\rangle\langle l|\partial_\nu H|n\rangle}{(E_n - E_m)(E_n - E_l)}|m\rangle - \sum_{m\neq n}\frac{\langle m|\partial_\mu H|n\rangle\langle n|\partial_\nu H|n\rangle}{(E_n - E_m)^2}|m\rangle - \frac{1}{2}\sum_{m\neq n}\frac{\langle n|\partial_\mu H|m\rangle\langle m|\partial_\nu H|n\rangle}{(E_n - E_m)^2}|m\rangle\right)x^\mu x^\nu + \cdots.\end{aligned}$$

Both energy derivatives and state derivatives will be involved in deduction. Whenever a state derivative is encountered, resolve it by inserting the complete set of basis, then the Hellmann-Feynman theorem is applicable. Because differentiation can be calculated systematically, the series expansion approach to the perturbative corrections can be coded on computers with symbolic processing software like Mathematica.

Effective Hamiltonian

Let $H(0)$ be the Hamiltonian completely restricted either in the low-energy subspace $\mathcal{H}_L$ or in the high-energy subspace $\mathcal{H}_H$, such that there is no matrix element in $H(0)$ connecting the low- and the high-energy subspaces, i.e. $\langle m | H(0) | l \rangle = 0$ if $m \in \mathcal{H}_L, l \in \mathcal{H}_H$. Let $F_\mu = \partial_\mu H$ be the coupling terms connecting the subspaces. Then when the high energy degrees of freedoms are integrated out, the effective Hamiltonian in the low energy subspace reads:

$$H_{mn}^{\text{eff}}(x^\mu) = \langle m | H | n \rangle + \delta_{nm} \langle m | \partial_\mu H | n \rangle x^\mu + \frac{1}{2!} \sum_{l \in \mathcal{H}_H} \left(\frac{\langle m | \partial_\mu H | l \rangle \langle l | \partial_\nu H | n \rangle}{E_m - E_l} + \frac{\langle m | \partial_\nu H | l \rangle \langle l | \partial_\mu H | n \rangle}{E_n - E_l} \right) x^\mu x^\nu + \cdots .$$

Here $m, n \in \mathcal{H}_L$ are restricted in the low energy subspace. The above result can be derived by power series expansion of $\langle m | H(x^\mu) | n \rangle$.

In a formal way it is possible to define an effective Hamiltonian that gives exactly the low-lying energy states and wavefunctions. In practice, some kind of approximation (perturbation theory) is generally required.

Time-dependent Perturbation Theory

Method of Variation of Constants

Time-dependent perturbation theory, developed by Paul Dirac, studies the effect of a time-dependent perturbation $V(t)$ applied to a time-independent Hamiltonian H_0.

Since the perturbed Hamiltonian is time-dependent, so are its energy levels and eigenstates. Thus, the goals of time-dependent perturbation theory are slightly different from time-independent perturbation theory. One is interested in the following quantities:

- The time-dependent expectation value of some observable A, for a given initial state.
- The time-dependent amplitudes of those quantum states that are energy eigenkets (eigenvectors) in the unperturbed system.

The first quantity is important because it gives rise to the classical result of an A measurement performed on a macroscopic number of copies of the perturbed system. For example, we could take A to be the displacement in the x-direction of the electron in a hydrogen atom, in which case the expected value, when multiplied by an appropriate coefficient, gives the time-dependent dielectric polarization of a hydrogen gas. With an appropriate choice of perturbation (i.e. an oscillating electric potential), this allows one to calculate the AC permittivity of the gas.

The second quantity looks at the time-dependent probability of occupation for each eigenstate. This is particularly useful in laser physics, where one is interested in the populations of different atomic states in a gas when a time-dependent electric field is applied. These probabilities are also useful for calculating the "quantum broadening" of spectral lines and particle decay in particle physics and nuclear physics.

We will briefly examine the method behind Dirac's formulation of time-dependent perturbation theory. Choose an energy basis $|n\rangle$ for the unperturbed system. (We drop the (0) superscripts for the eigenstates, because it is not useful to speak of energy levels and eigenstates for the perturbed system).

If the unperturbed system is an eigenstate (of the Hamiltonian) $|j\rangle$ at time $t = 0$, its state at subsequent times varies only by a phase (in the Schrödinger picture, where state vectors evolve in time and operators are constant),

$$|j(t)\rangle = e^{-iE_j t/\hbar}|j\rangle.$$

Now, introduce a time-dependent perturbing Hamiltonian $V(t)$. The Hamiltonian of the perturbed system is:

$$H = H_0 + V(t).$$

Let $|\psi(t)\rangle$ denote the quantum state of the perturbed system at time t. It obeys the time-dependent Schrödinger equation,

$$H|\psi(t)\rangle = i\hbar\frac{\partial}{\partial t}|\psi(t)\rangle.$$

The quantum state at each instant can be expressed as a linear combination of the complete eigenbasis of $|n\rangle$:

$$|\psi(t)\rangle = \sum_n c_n(t)e^{-iE_n t/\hbar}|n\rangle,$$

where the $c_n(t)$s are to be determined complex functions of t which we will refer to as amplitudes (strictly speaking, they are the amplitudes in the Dirac picture).

We have explicitly extracted the exponential phase factors $\exp(-iE_n t/\hbar)$ on the right hand side. This is only a matter of convention, and may be done without loss of generality. The reason we go to this trouble is that when the system starts in the state $|j\rangle$ and no perturbation is present, the amplitudes have the convenient property that, for all t, $c_j(t) = 1$ and $c_n(t) = 0$ if $n \neq j$.

The square of the absolute amplitude $c_n(t)$ is the probability that the system is in state n at time t, since:

$$|c_n(t)|^2 = |\langle n|\psi(t)\rangle|^2.$$

Plugging into the Schrödinger equation and using the fact that ∂/∂t acts by a product rule, one obtains:

$$\sum_n \left(i\hbar \frac{dc_n}{dt} - c_n(t)V(t) \right) e^{-iE_n t/\hbar} \mid n\rangle = 0.$$

By resolving the identity in front of V and multiplying through by the bra $\langle n \mid$ on the left, this can be reduced to a set of coupled differential equations for the amplitudes,

$$\frac{dc_n}{dt} = \frac{-i}{\hbar} \sum_k \langle n \mid V(t) \mid k\rangle c_k(t) e^{-i(E_k - E_n)t/\hbar}.$$

where we have used equation to evaluate the sum on n in the second term, then used the fact that $\langle k \mid \Psi(t)\rangle = c_k(t)e^{-iE_k t/\hbar}$.

The matrix elements of V play a similar role as in time-independent perturbation theory, being proportional to the rate at which amplitudes are shifted between states. Note, however, that the direction of the shift is modified by the exponential phase factor. Over times much longer than the energy difference $E_k - E_n$, the phase winds around o several times. If the time-dependence of V is sufficiently slow, this may cause the state amplitudes to oscillate. (E.g., such oscillations are useful for managing radiative transitions in a laser).

Up to this point, we have made no approximations, so this set of differential equations is exact. By supplying appropriate initial values $c_n(t)$, we could in principle find an exact (i.e., non-perturbative) solution. This is easily done when there are only two energy levels (n = 1, 2), and this solution is useful for modelling systems like the ammonia molecule.

However, exact solutions are difficult to find when there are many energy levels, and one instead looks for perturbative solutions. These may be obtained by expressing the equations in an integral form,

$$c_n(t) = c_n(0) + \frac{-i}{\hbar} \sum_k \int_0^t dt' \langle n \mid V(t') \mid k\rangle c_k(t') e^{-i(E_k - E_n)t'/\hbar}.$$

Repeatedly substituting this expression for c_n back into right hand side, yields an iterative solution,

$$c_n(t) = c_n^{(0)} + c_n^{(1)} + c_n^{(2)} + \cdots$$

where, for example, the first-order term is:

$$c_n^{(1)}(t) = \frac{-i}{\hbar} \sum_k \int_0^t dt' \langle n \mid V(t') \mid k\rangle c_k^{(0)} e^{-i(E_k - E_n)t'/\hbar}.$$

Several further results follow from this, such as Fermi's golden rule, which relates the rate of transitions between quantum states to the density of states at particular energies; or the Dyson series, obtained by applying the iterative method to the time evolution operator, which is one of the starting points for the method of Feynman diagrams.

Method of Dyson Series

Time-dependent perturbations can be reorganized through the technique of the Dyson series. The Schrödinger equation:

$$H(t)|\psi(t)\rangle = i\hbar\frac{\partial|\psi(t)\rangle}{\partial t}$$

has the formal solution:

$$|\psi(t)\rangle = T\exp\left[-\frac{i}{\hbar}\int_{t_0}^{t} dt' H(t')\right]|\psi(t_0)\rangle,$$

where T is the time ordering operator,

$$TA(t_1)A(t_2) = \begin{cases} A(t_1)A(t_2) & t_1 > t_2 \\ A(t_2)A(t_1) & t_2 > t_1 \end{cases}.$$

Thus, the exponential represents the following Dyson series,

$$|\psi(t)\rangle = \left[1 - \frac{i}{\hbar}\int_{t_0}^{t} dt_1 H(t_1) - \frac{1}{\hbar^2}\int_{t_0}^{t} dt_1 \int_{t_0}^{t_1} dt_2 H(t_1)H(t_2) + \ldots\right]|\psi(t_0)\rangle.$$

In the second term, the 1/2! factor exactly cancels the double contribution due to the time-ordering operator, etc.

Consider the following perturbation problem:

$$[H_0 + \lambda V(t)]|\psi(t)\rangle = i\hbar\frac{\partial|\psi(t)\rangle}{\partial t},$$

assuming that the parameter λ is small and that the problem $H_0|n\rangle = E_n|n\rangle$ has been solved.

Perform the following unitary transformation to the interaction picture (or Dirac picture),

$$|\psi(t)\rangle = e^{-\frac{i}{\hbar}H_0(t-t_0)}|\psi_I(t)\rangle.$$

Consequently, the Schrödinger equation simplifies to:

$$\lambda e^{\frac{i}{\hbar}H_0(t-t_0)}V(t)e^{-\frac{i}{\hbar}H_0(t-t_0)}|\psi_I(t)\rangle = i\hbar\frac{\partial|\psi_I(t)\rangle}{\partial t},$$

so it is solved through the above Dyson series,

$$\psi_I(t)\rangle = \left[\begin{array}{l} 1-\frac{i\lambda}{\hbar}\int_{t_0}^{t} dt_1 e^{\frac{i}{\hbar}H_0(t_1-t_0)}V(t_1)e^{-\frac{i}{\hbar}H_0(t_1-t_0)} \\ -\frac{\lambda^2}{\hbar^2}\int_{t_0}^{t} dt_1\int_{t_0}^{t_1} dt_2 e^{\frac{i}{\hbar}H_0(t_1-t_0)}V(t_1)e^{-\frac{i}{\hbar}H_0(t_1-t_0)}e^{\frac{i}{\hbar}H_0(t_2-t_0)}V(t_2)e^{-\frac{i}{\hbar}H_0(t_2-t_0)}+\ldots \end{array}\right]|\psi(t_0)\rangle,,$$

as a perturbation series with small λ.

Using the solution of the unperturbed problem $H_0|n\rangle = E_n|n\rangle$ and $\sum_n |n\rangle\langle n| = 1$ (for the sake of simplicity assume a pure discrete spectrum), yields, to first order,

$$|\psi_I(t)\rangle = \left[1-\frac{i\lambda}{\hbar}\sum_m\sum_n\int_{t_0}^{t} dt_1\langle m|V(t_1)|n\rangle e^{-\frac{i}{\hbar}(E_n-E_m)(t_1-t_0)}|m\rangle\langle n|+\ldots\right]|\psi(t_0)\rangle.$$

Thus, the system, initially in the unperturbed state $|\alpha\rangle = |\psi(t_0)\rangle$, by dint of the perturbation can go into the state $|\beta\rangle$. The corresponding transition probability amplitude to first order is:

$$A_{\alpha\beta} = -\frac{i\lambda}{\hbar}\int_{t_0}^{t} dt_1\langle\beta|V(t_1)|\alpha\rangle e^{-\frac{i}{\hbar}(E_\alpha-E_\beta)(t_1-t_0)},$$

while the corresponding transition probability to a continuum is furnished by Fermi's golden rule.

As an aside, note that time-independent perturbation theory is also organized inside this time-dependent perturbation theory Dyson series. To see this, write the unitary evolution operator, obtained from the above Dyson series, as:

$$\begin{aligned} U(t) = 1&-\frac{i\lambda}{\hbar}\int_{t_0}^{t} dt_1 e^{\frac{i}{\hbar}H_0(t_1-t_0)}V(t_1)e^{-\frac{i}{\hbar}H_0(t_1-t_0)} \\ &-\frac{\lambda^2}{\hbar^2}\int_{t_0}^{t} dt_1\int_{t_0}^{t_1} dt_2 e^{\frac{i}{\hbar}H_0(t_1-t_0)}V(t_1)e^{-\frac{i}{\hbar}H_0(t_1-t_0)}e^{\frac{i}{\hbar}H_0(t_2-t_0)}V(t_2)e^{-\frac{i}{\hbar}H_0(t_2-t_0)}+\cdots \end{aligned}$$

and take the perturbation V to be time-independent.

Using the identity resolution:

$$\sum_n |n\rangle\langle n| = 1$$

with $H_0 \mid n\rangle = E_n \mid n\rangle$ for a pure discrete spectrum, write:

$$U(t) = 1 - \left\{ \frac{i\lambda}{\hbar} \int_{t_0}^{t} dt_1 \sum_m \sum_n \langle m \mid V \mid n\rangle e^{-\frac{i}{\hbar}(E_n - E_m)(t_1 - t_0)} \mid m\rangle\langle n \mid \right\}$$

$$- \left\{ \frac{\lambda^2}{\hbar^2} \int_{t_0}^{t} dt_1 \int_{t_0}^{t_1} dt_2 \sum_m \sum_n \sum_q e^{-\frac{i}{\hbar}(E_n - E_m)(t_1 - t_0)} \langle m \mid V \mid n\rangle\langle n \mid V \mid q\rangle e^{-\frac{i}{\hbar}(E_q - E_n)(t_2 - t_0)} \mid m\rangle\langle q \mid \right\} + \cdots$$

It is evident that, at second order, one must sum on all the intermediate states. Assume $t_0 = 0$ and the asymptotic limit of larger times. This means that, at each contribution of the perturbation series, one has to add a multiplicative factor $e^{-\epsilon t}$ in the integrands for ε arbitrarily small. Thus the limit $t \to \infty$ gives back the final state of the system by eliminating all oscillating terms, but keeping the secular ones. The integrals are thus computable, and, separating the diagonal terms from the others yields:

$$U(t) = 1 - \frac{i\lambda}{\hbar} \sum_n \langle n \mid V \mid n\rangle t - \frac{i\lambda^2}{\hbar} \sum_{m \neq n} \frac{\langle n \mid V \mid m\rangle\langle m \mid V \mid n\rangle}{E_n - E_m} t - \frac{1}{2} \frac{\lambda^2}{\hbar^2} \sum_{m,n} \langle n \mid V \mid m\rangle\langle m \mid V \mid n\rangle t^2 + \ldots$$

$$+ \lambda \sum_{m \neq n} \frac{\langle m \mid V \mid n\rangle}{E_n - E_m} \mid m\rangle\langle n \mid + \lambda^2 \sum_{m \neq n} \sum_{q \neq n} \sum_n \frac{\langle m \mid V \mid n\rangle\langle n \mid V \mid q\rangle}{(E_n - E_m)(E_q - E_n)} \mid m\rangle\langle q \mid + \ldots$$

where the time secular series yields the eigenvalues of the perturbed problem specified above, recursively; whereas the remaining time-constant part yields the corrections to the stationary eigenfunctions also given above ($\mid n(\lambda)\rangle = U(o;\lambda) \mid n\rangle$).

The unitary evolution operator is applicable to arbitrary eigenstates of the unperturbed problem and, in this case, yields a secular series that holds at small times.

Strong Perturbation Theory

In a similar way as for small perturbations, it is possible to develop a strong perturbation theory. Consider as usual the Schrödinger equation:

$$H(t) \mid \psi(t)\rangle = i\hbar \frac{\partial \mid \psi(t)\rangle}{\partial t}$$

and we consider the question if a dual Dyson series exists that applies in the limit of a perturbation increasingly large. This question can be answered in an affirmative way and the series is the well-known adiabatic series. This approach is quite general and can be shown in the following way. Consider the perturbation problem:

$$[H_0 + \lambda V(t)] \mid \psi(t)\rangle = i\hbar \frac{\partial \mid \psi(t)\rangle}{\partial t}$$

being $\lambda \to \infty$. Our aim is to find a solution in the form:

$$\mid \psi\rangle = \mid \psi_0\rangle + \frac{1}{\lambda} \mid \psi_1\rangle + \frac{1}{\lambda^2} \mid \psi_2\rangle + \ldots$$

but a direct substitution into the above equation fails to produce useful results. This situation can be adjusted making a rescaling of the time variable as $\tau = \lambda t$ producing the following meaningful equations:

$$V(t)|\psi_0\rangle = i\hbar\frac{\partial|\psi_0\rangle}{\partial\tau}$$

$$V(t)|\psi_1\rangle + H_0|\psi_0\rangle = i\hbar\frac{\partial|\psi_1\rangle}{\partial\tau}...$$

that can be solved once we know the solution of the leading order equation. But we know that in this case we can use the adiabatic approximation. When V (t) does not depend on time one gets the Wigner-Kirkwood series that is often used in statistical mechanics. Indeed, in this case we introduce the unitary transformation:

$$|\psi(t)\rangle = e^{-\frac{i}{\hbar}\lambda V(t-t_0)}|\psi_F(t)\rangle$$

that defines a free picture as we are trying to eliminate the interaction term. Now, in dual way with respect to the small perturbations, we have to solve the Schrödinger equation:

$$e^{\frac{i}{\hbar}\lambda V(t-t_0)}H_0 e^{-\frac{i}{\hbar}\lambda V(t-t_0)}|\psi_F(t)\rangle = i\hbar\frac{\partial|\psi_F(t)\rangle}{\partial t}$$

and we see that the expansion parameter λ appears only into the exponential and so, the corresponding Dyson series, a dual Dyson series, is meaningful at large λs and is:

$$|\psi_F(t)\rangle = \left[\begin{array}{l} 1-\frac{i}{\hbar}\int_{t_0}^{t} dt_1 e^{\frac{i}{\hbar}\lambda V(t_1-t_0)}H_0 e^{-\frac{i}{\hbar}\lambda V(t_1-t_0)} \\ -\frac{1}{\hbar^2}\int_{t_0}^{t} dt_1\int_{t_0}^{t_1} dt_2 e^{\frac{i}{\hbar}\lambda V(t_1-t_0)}H_0 e^{-\frac{i}{\hbar}\lambda V(t_1-t_0)} e^{\frac{i}{\hbar}\lambda V(t_2-t_0)}H_0 e^{-\frac{i}{\hbar}\lambda V(t_2-t_0)} + \ldots \end{array}\right]|\psi(t_0)\rangle.$$

After the rescaling in time $\tau = \lambda t$ we can see that this is indeed a series in $1/\lambda$ justifying in this way the name of dual Dyson series. The reason is that we have obtained this series simply interchanging H_0 and V and we can go from one to another applying this exchange. This is called duality principle in perturbation theory. The choice $H_0 = p^2/2m$ yields, as already said, a Wigner-Kirkwood series that is a gradient expansion. The Wigner-Kirkwood series is a semiclassical series with eigenvalues given exactly as for WKB approximation.

Example of first order perturbation theory – ground state energy of the quartic oscillator. Consider the quantum harmonic oscillator with the quartic potential perturbation and the Hamiltonian:

$$H = -\frac{\hbar^2}{2m}\frac{\partial^2}{\partial x^2} + \frac{m\omega^2 x^2}{2} + \lambda x^4$$

The ground state of the harmonic oscillator is:

$$\psi_0 = \left(\frac{\alpha}{\pi}\right)^{\frac{1}{4}} e^{-\alpha x^2/2}$$

($\alpha = m\omega / \hbar$) and the energy of unperturbed ground state is:

$$E_0^{(0)} = \tfrac{1}{2}\hbar\omega.$$

Using the first order correction formula we get:

$$E_0^{(1)} = \lambda\left(\frac{\alpha}{\pi}\right)^{\frac{1}{2}} \int e^{-\alpha x^2/2} x^4 e^{-\alpha x^2/2} dx = \lambda\left(\frac{\alpha}{\pi}\right)^{\frac{1}{2}} \frac{\partial^2}{\partial \alpha^2} \int e^{-\alpha x^2} dx$$

or:

$$E_0^{(1)} = \lambda\left(\frac{\alpha}{\pi}\right)^{\frac{1}{2}} \frac{\partial^2}{\partial \alpha^2}\left(\frac{\pi}{\alpha}\right)^{\frac{1}{2}} = \lambda\frac{3}{4}\frac{1}{\alpha^2} = \frac{3}{4}\frac{\hbar^2\lambda}{m^2\omega^2}$$

Example of first and second order perturbation theory – quantum pendulum. Consider the quantum mathematical pendulum with the Hamiltonian:

$$H = -\frac{\hbar^2}{2ma^2}\frac{\partial^2}{\partial \phi^2} - \lambda\cos\phi$$

with the potential energy $-\lambda\cos\phi$ taken as the perturbation i.e.,

$$V = -\cos\phi$$

The unperturbed normalized quantum wave functions are those of the rigid rotor and are given by:

$$\psi_n(\phi) = \frac{e^{in\phi}}{\sqrt{2\pi}}$$

and the energies:

$$E_n^{(0)} = \frac{\hbar^2 n^2}{2ma^2}$$

The first order energy correction to the rotor due to the potential energy is:

$$E_n^{(1)} = -\frac{1}{2\pi}\int e^{-in\phi}\cos\phi e^{in\phi} = -\frac{1}{2\pi}\int \cos\phi = 0$$

Using the formula for the second order correction one gets:

$$E_n^{(2)} = \frac{ma^2}{2\pi^2\hbar^2}\sum_k \frac{\left|\int e^{-ik\phi}\cos\phi e^{in\phi}\right|^2}{n^2-k^2}$$

or,

$$E_n^{(2)} = \frac{ma^2}{2\hbar^2}\sum_k \frac{\left|\left(\delta_{n,1-k}+\delta_{n,-1-k}\right)\right|^2}{n^2-k^2}$$

or,

$$E_n^{(2)} = \frac{ma^2}{2\hbar^2}\left(\frac{1}{2n-1}+\frac{1}{-2n-1}\right) = \frac{ma^2}{\hbar^2}\frac{1}{4n^2-1}.$$

SCATTERING THEORY

Quantum particles exhibit a feature known as wave-particle duality, which can be summarized in the quantum double-slit thought experiment. As shown in the figure below, a source emits electrons with energy E, which travel towards a screen with a pair of slits. A detector is positioned on the other side of the screen. By moving the detector around, we can measure the rate at which electrons are detected at different positions.

According to quantum theory, the experiment reveals the following: (i) the electrons arrive in discrete units—one at a time, like classical particles; (ii) when we move the detector around to measure how the detection events are statistically distributed in space, the resulting distribution matches an interference pattern formed by a classical wave diffracted by the slits. The wavelength λ is related to the electron energy E by:

$$\lambda = \frac{2\pi}{k},\ E = \frac{\hbar^2 k^2}{2m},$$

where $\hbar = h/2\pi$ is Dirac's constant, and m is the electron mass. (These relations imply, by the way, that we can deduce the spacing of the slits from the diffraction pattern, if the incident energy E is known).

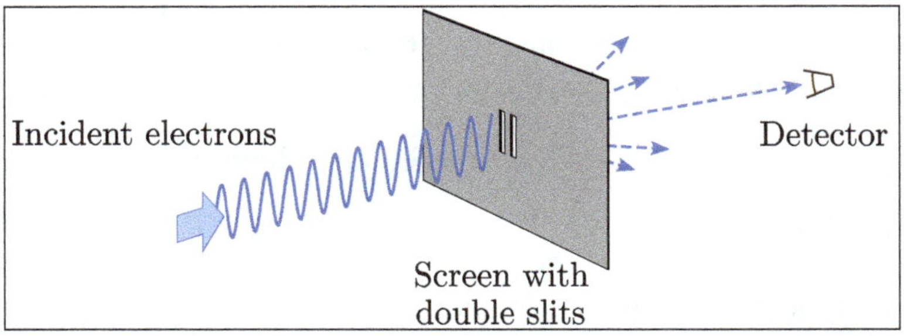

Wave-particle duality arises from quantum theory's distinction between a particle's state and the outcomes of measurements performed on it. The state is described by a wavefunction $\psi(\mathbf{r})$, which can undergo diffraction like a classical wave. Measurement outcomes, however, depend probabilistically on the wavefunction. In a position measurement, the probability of locating a particle in a volume dV around position r is $|\psi(\mathbf{r})|^2 \, dV$.

We will study a generalization of the double-slit experiment called a scattering experiment. The idea is to take an object called a scatterer, shoot quantum particles at it, and measure the resulting particle distribution. Just as the double-slit interference pattern can be used to deduce the slit spacing, a scattering experiment can be used to deduce various facts about the scatterer. Scattering experiments, as a class, constitute a large proportion of the methods used to probe the quantum world—from electronand photon-based laboratory experiments for measuring the properties of materials, to huge accelerator experiments that study high-energy phenomena like the Higgs boson.

We will focus on a relatively simple scenario, consisting of a single non-relativistic quantum particle and a classical scatterer. Consider a continuous and unbounded d-dimensional space, describable by coordinates **r**. Somewhere around the origin, **r** = 0, is a finite-sized scatterer. An incoming quantum particle, with energy E, is governed by the Hamiltonian:

$$\hat{H} = \hat{H}_0 + V(\hat{r}), \ \hat{H}_0 = \frac{\hat{P}^2}{2m}.$$

Here, $\hat{H}$ 0 describes the particle's kinetic energy, m is the particle's mass, $\hat{r}$ and $\hat{p}$ are position and momentum operators, and V is a scattering potential describing how the scatterer affects the quantum particle. We assume that V (**r**) → 0 as |r| → ∞, i.e., the scattering potential becomes negligible far from the origin.

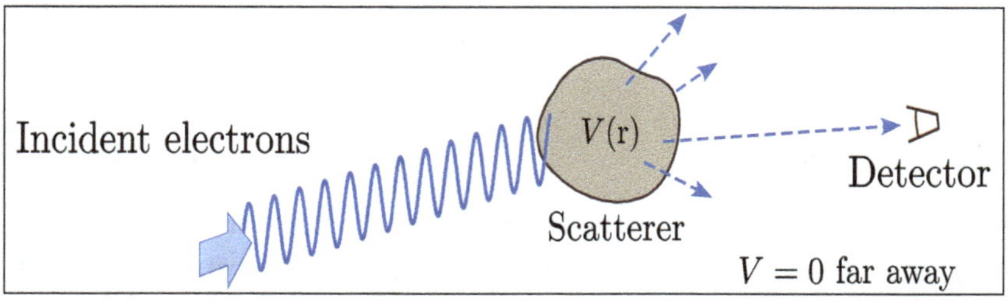

We prepare an incoming particle state with energy E, and want to see how the particle is scattered by the potential. However, converting these words into a well-defined mathematical problem is a bit tricky. We will give the formulation first, before discussing its meaning:

- The particle state $|\psi\rangle$ obeys the time-independent Schrödinger equation,

  $$\hat{H}|\psi\rangle = E|\psi\rangle$$

  where E is the incoming particle energy.

- This state can be decomposed into two terms,

  $|\psi\rangle = |\psi_i\rangle + |\psi_s\rangle,$

  where $|\psi_i\rangle$ is called the incident state and $|\psi_s\rangle$ is called the scattered state.
- The incident state is an eigenstate of $\hat{H}_0$ with energy E,

  $\hat{H}_0|\psi\rangle = E|\psi_i\rangle.$
- Lastly, we require the scattered state to be an "outgoing" state. This is the most subtle requirement, and we will describe how to deal with it later.

The first condition says that the scattering process is elastic; since the scatterer takes the form of a potential V (**r**), its interaction with the particle is conservative (i.e., the total energy E is fixed). The second condition, viewed in the position basis, says that the particle's wavefunction ψ(**r**) = $\langle r|\psi\rangle$ can be taken as a superposition of an incoming wave and a scattered wave. The third condition says that far from the scatterer (|r| → ∞), the incident wave has a fixed wavelength determined by the energy E. The final condition says that the scattered part of the wavefunction must correspond to a wave moving out to infinity.

Given $|\psi_i\rangle$, E, and V (**r**), we are to solve for $|\psi_s\rangle$. This is not an eigenproblem. Usually, when dealing with the time-independent Schrödinger equation, we treat it as an eigenproblem and solve for the energy eigenvalues and eigenstates. Here, however, E is not an output of the calculation, but part of the input–it is a parameter describing the energy of the incoming particles in the experiment.

SCATTERING LENGTH

The scattering length in quantum mechanics describes low-energy scattering. For potentials that decay faster than $1/r^3 1/r^3 asr \to \infty r \to \infty$, it is defined as the following low-energy limit:

$$\lim_{k\to 0} k \cot \delta(k) = -\frac{1}{a},$$

where a is the scattering length, k is the wave number, and $\delta(k)$ is the phase shift of the outgoing spherical wave. The elastic cross section, σ_e, at low energies is determined solely by the scattering length.

When a slow particle scatters off a short ranged scatterer (e.g. an impurity in a solid or a heavy particle) it cannot resolve the structure of the object since its de Broglie wavelength is very long. The idea is that then it should not be important what precise potential V(r) one scatters off, but only how the potential looks at long length scales. The formal way to solve this problem is to do a partial wave expansion (somewhat analogous to the multipole expansion in classical electrodynamics), where one expands in the angular momentum components of the outgoing wave. At very low energy the incoming particle does not see any structure, therefore to lowest order one has only a spherical outgoing wave, called the s-wave in analogy with the atomic orbital at angular momentum quantum number l=0. At higher energies one also needs to consider p and d-wave (l = 1, 2) scattering and so on.

The idea of describing low energy properties in terms of a few parameters and symmetries is very powerful, and is also behind the concept of renormalization.

The concept of the scattering length can also be extended to potentials that decay slower than $1/r^3$ as $r \to \infty$. A famous example, relevant for proton-proton scattering, is the Coulomb-modified scattering length.

As an example on how to compute the s-wave (i.e. angular momentum $l = 0$) scattering length for a given potential we look at the infinitely repulsive spherical potential well of radius r_0 in 3 dimensions. The radial Schrödinger equation ($l = 0$) outside of the well is just the same as for a free particle:

$$-\frac{\hbar^2}{2m}u''(r) = Eu(r),$$

where the hard core potential requires that the wave function $u(r)$ vanishes at $r = r_0$, $u(r_0) = 0$. The solution is readily found:

$$u(r) = A\sin(kr + \delta_s).$$

Here $k = \sqrt{2mE}/\hbar$ and $\delta_s = -k \cdot r_0$ is the s-wave phase shift (the phase difference between incoming and outgoing wave), which is fixed by the boundary condition $u(r_0) = 0$; A is an arbitrary normalization constant.

One can show that in general $\delta_s(k) \approx -k \cdot a_s + O(k^2)$ for small k (i.e. low energy scattering). The parameter a_s of dimension length is defined as the scattering length. For our potential we have therefore $a = r_0$, in other words the scattering length for a hard sphere is just the radius. (Alternatively one could say that an arbitrary potential with s-wave scattering length a_s has the same low energy scattering properties as a hard sphere of radius a_s). To relate the scattering length to physical observables that can be measured in a scattering experiment we need to compute the cross section σ. In scattering theory one writes the asymptotic wavefunction as (we assume there is a finite ranged scatterer at the origin and there is an incoming plane wave along the z-axis):

$$\psi(r,\theta) = e^{ikz} + f(\theta)\frac{e^{ikr}}{r}$$

where f is the scattering amplitude. According to the probability interpretation of quantum mechanics the differential cross section is given by $d\sigma / d\Omega = |f(\theta)|^2$ (the probability per unit time to scatter into the direction k). If we consider only s-wave scattering the differential cross section does not depend on the angle θ, and the total scattering cross section is just $\sigma = 4\pi |f|^2$. The s-wave part of the wavefunction $\psi(r,\theta)$ is projected out by using the standard expansion of a plane wave in terms of spherical waves and Legendre polynomials $P_l(\cos\theta)$:

$$e^{ikz} \approx \frac{1}{2ikr} \sum_{l=0}^{\infty} (2l+1) P_l(\cos\theta) \left[(-1)^{l+1} e^{-ikr} + e^{ikr} \right]$$

Bymatchingthe $l = 0$ componentof $\psi(r,\theta)$ tothes-wavesolution $\psi(r) = A\sin(kr + \delta_s)/r$ (where we normalize A such that the incoming wave e^{ikz} has a prefactor of unity) one has:

$$f = \frac{1}{2ik}(e^{2i\delta_s} - 1) \approx \delta_s / k \approx -a_s$$

This gives:

$$\sigma = \frac{4\pi}{k^2} \sin^2 \delta_s = 4\pi a_s^2$$

S-MATRIX

In physics, the S-matrix or scattering matrix relates the initial state and the final state of a physical system undergoing a scattering process. It is used in quantum mechanics, scattering theory and quantum field theory (QFT).

More formally, in the context of QFT, the S-matrix is defined as the unitary matrix connecting sets of asymptotically free particle states (the *in-states* and the *out-states*) in the Hilbert space of physical states. A multi-particle state is said to be *free* (non-interacting) if it transforms under Lorentz transformations as a tensor product, or *direct product* in physics parlance, of *one-particle states* as prescribed by equation below,

$$U(\Lambda, a)\Psi_{p_1\sigma_1 n_1; p_2\sigma_2 n_2 \cdots} = e^{-ia_\mu((\Lambda p_1)^\mu + (\Lambda p_2)^\mu + \cdots)} \sqrt{\frac{(\Lambda p_1)^0 (\Lambda p_2)^0 \cdots}{p_1^0 p_2^0 \cdots}} \sum_{\sigma_1' \sigma_2' \cdots} D^{(j_1)}_{\sigma_1' \sigma_1}(W(\Lambda, p_1)) D^{(j_2)}_{\sigma_2' \sigma_2}(W(\Lambda, p_2)) \cdots \Psi_{\Lambda p_1 \sigma_1' n_1; \Lambda p_2 \sigma_2' n_2 \cdots},$$

asymptotically free then means that the state has this appearance in either the distant past or the distant future.

While the S-matrix may be defined for any background (spacetime) that is asymptotically solvable and has no event horizons, it has a simple form in the case of the

Minkowski space. In this special case, the Hilbert space is a space of irreducible unitary representations of the inhomogeneous Lorentz group (the Poincaré group); the S-matrix is the evolution operator between time equal to minus infinity (the distant past), and time equal to plus infinity (the distant future). It is defined only in the limit of zero energy density (or infinite particle separation distance).

It can be shown that if a quantum field theory in Minkowski space has a mass gap, the state in the asymptotic past and in the asymptotic future are both described by Fock spaces.

The S-matrix was first introduced by John Archibald Wheeler in the 1937 paper "On the Mathematical Description of Light Nuclei by the Method of Resonating Group Structure". In this paper Wheeler introduced a *scattering matrix* – a unitary matrix of coefficients connecting "the asymptotic behaviour of an arbitrary particular solution [of the integral equations] with that of solutions of a standard form", but did not develop it fully.

In the 1940s, Werner Heisenberg developed, independently, and substantiated the idea of the S-matrix. Because of the problematic divergences present in quantum field theory at that time, Heisenberg was motivated to isolate the *essential features of the theory* that would not be affected by future changes as the theory developed. In doing so, he was led to introduce a unitary "characteristic" S-matrix.

Today, however, exact S-matrix results are a crowning achievement of conformal field theory, integrable systems, and several further areas of quantum field theory and string theory. S-matrices are not substitutes for a field-theoretic treatment, but rather, complement the end results of such.

The S-matrix is closely related to the transition probability amplitude in quantum mechanics and to cross sections of various interactions; the elements (individual numerical entries) in the S-matrix are known as scattering amplitudes. Poles of the S-matrix in the complex-energy plane are identified with bound states, virtual states or resonances. Branch cuts of the S-matrix in the complex-energy plane are associated to the opening of a scattering channel.

In the Hamiltonian approach to quantum field theory, the S-matrix may be calculated as a time-ordered exponential of the integrated Hamiltonian in the interaction picture; it may also be expressed using Feynman's path integrals. In both cases, the perturbative calculation of the S-matrix leads to Feynman diagrams.

In scattering theory, the S-matrix is an operator mapping free particle *in-states* to free particle *out-states* (scattering channels) in the Heisenberg picture. This is very useful because often we cannot describe the interaction (at least, not the most interesting ones) exactly.

In One-dimensional Quantum Mechanics

A simple prototype in which the S-matrix is 2-dimensional is considered first, for the purposes of illustration. In it, particles with sharp energy E scatter from a localized potential V according to the rules of 1-dimensional quantum mechanics. Already this simple model displays some features of more general cases, but is easier to handle.

Each energy E yields an S-matrix $S = S(E)$ that depends on V. Thus, the total S-matrix could, figuratively speaking, be visualized, in a suitable basis, as a "continuous matrix" with every element zero except for 2×2-blocks along the diagonal for a given V.

Consider a localized one dimensional potential barrier $V(x)$, subjected to a beam of quantum particles with energy E. These particles are incident on the potential barrier from left to right.

The solution of Schrödinger's equation outside the potential barrier are plane waves given by:

$$\psi_L(x) = Ae^{ikx} + Be^{-ikx}$$

for the region to the left of the potential barrier, and

$$\psi_R(x) = Ce^{ikx} + De^{-ikx}$$

for the region to the right to the potential barrier, where,

$$k = \sqrt{2mE/\hbar^2}$$

is the wave vector. The time dependence is not needed in our overview and is hence omitted. The term with coefficient A represents the incoming wave, whereas term with coefficient C represents the outgoing wave. B stands for the reflecting wave. Since we set the incoming wave moving in the positive direction (coming from the left), D is zero and can be omitted.

The "scattering amplitude", i.e., the transition overlap of the outgoing waves with the incoming waves is a linear relation defining the S-matrix,

$$\begin{pmatrix} B \\ C \end{pmatrix} = \begin{pmatrix} S_{11} & S_{12} \\ S_{21} & S_{22} \end{pmatrix} \begin{pmatrix} A \\ D \end{pmatrix}.$$

The above relation can be written as:

$$\Psi_{out} = S\Psi_{in}$$

where,

$$\Psi_{out} = \begin{pmatrix} B \\ C \end{pmatrix}, \quad \Psi_{in} = \begin{pmatrix} A \\ D \end{pmatrix}, \qquad S = \begin{pmatrix} S_{11} & S_{12} \\ S_{21} & S_{22} \end{pmatrix}.$$

The elements of S completely characterize the scattering properties of the potential barrier V(x).

Unitary Property

The unitary property of S-matrix is directly related to the conservation of the probability current in quantum mechanics.

The probability current J of the wave function $\psi(x)$ is defined as,

$$J = \frac{\hbar}{2mi}\left(\psi^* \frac{\partial \psi}{\partial x} - \psi \frac{\partial \psi^*}{\partial x}\right).$$

The current density to the left of the barrier is,

$$J_L = \frac{\hbar k}{m}\left(|A|^2 - |B|^2\right),$$

while the current density to the right of the barrier is,

$$J_R = \frac{\hbar k}{m}\left(|C|^2 - |D|^2\right).$$

For conservation of the probability current density, $J_L = J_R$. This implies the S-matrix is a unitary matrix:

$$\begin{aligned}
&J_L = J_R \\
&|A|^2 - |B|^2 = |C|^2 - |D|^2 \\
&|B|^2 + |C|^2 = |A|^2 + |D|^2 \\
&\Psi_{out}^\dagger \Psi_{out} = \Psi_{in}^\dagger \Psi_{in} \\
&\Psi_{in}^\dagger S^\dagger S \Psi_{in} = \Psi_{in}^\dagger \Psi_{in} \\
&S^\dagger S = I.
\end{aligned}$$

Time-reversal Symmetry

If the potential $V(x)$ is real, then the system possesses time-reversal symmetry. Under this condition, if $\psi(x)$ is a solution of Schrödinger's equation, then $\psi^*(x)$ is also a solution.

The time-reversed solution is given by:

$$\psi_L^*(x) = A^* e^{-ikx} + B^* e^{ikx}$$

for the region to the left to the potential barrier, and,

$$\psi_R^*(x) = C^* e^{-ikx} + D^* e^{ikx}$$

for the region to the right to the potential barrier, where the terms with coefficient B^*, C^* represent incoming wave, and terms with coefficient A^*, D^* represent outgoing wave.

They are again related by the S-matrix,

$$\begin{pmatrix} A^* \\ D^* \end{pmatrix} = \begin{pmatrix} S_{11} & S_{12} \\ S_{21} & S_{22} \end{pmatrix} \begin{pmatrix} B^* \\ C^* \end{pmatrix}$$

that is,

$$\Psi_{in}^* = S\Psi_{out}^*.$$

Now, the relations,

$$\Psi_{in}^* = S\Psi_{out}^*, \quad \Psi_{out} = S\Psi_{in}$$

together yield a condition,

$$S^* S = I$$

This condition, in conjunction with the unitarity relation, implies that the S-matrix is symmetric, as a result of time reversal symmetry,

$$S^T = S.$$

Transmission Coefficient and Reflection Coefficient

The transmission coefficient from the left of the potential barrier is, when $D = 0$,

$$T_L = \frac{|C|^2}{|A|^2} = |S_{21}|^2.$$

The reflection coefficient from the left of the potential barrier is, when D = 0,

$$R_L = \frac{|B|^2}{|A|^2} = |S_{11}|^2.$$

Similarly, the transmission coefficient from the right of the potential barrier is, when A = 0,

$$T_R = \frac{|B|^2}{|D|^2} = |S_{12}|^2.$$

The reflection coefficient from the right of the potential barrier is, when A = 0,

$$R_R = \frac{|C|^2}{|D|^2} = |S_{22}|^2.$$

The relations between the transmission and reflection coefficients are,

$$T_L + R_L = 1$$

and,

$$T_R + R_R = 1.$$

This is a consequence of the unitarity property of the S-matrix.

Optical Theorem in One Dimension

In the case of free particles $V(x) = 0$, the S-matrix is:

$$S = \begin{pmatrix} 0 & 1 \\ 1 & 0 \end{pmatrix}.$$

Whenever V(x) is different from zero, however, there is a departure of the S-matrix from the above form, to,

$$S = \begin{pmatrix} 2ir & 1+2it \\ 1+2it & 2ir^* \dfrac{1+2it}{1-2it^*} \end{pmatrix}.$$

This departure is parameterized by two complex functions of energy, r and t. From unitarity there also follows a relationship between these two functions,

$$|r|^2 + |t|^2 = \mathrm{Im}(t).$$

The analogue of this identity in three dimensions is known as the optical theorem.

Quantum Field Theory

Interaction Picture

A straightforward way to define the S-matrix begins with considering the interaction picture. Let the Hamiltonian H be split into the free part H_0 and the interaction V, $H = H_0 + V$. In this picture, the operators behave as free field operators and the state vectors have dynamics according to the interaction V. Let,

$$|\Psi(t)\rangle$$

denote a state that has evolved from a free initial state,

$$|\Phi_i\rangle.$$

The S-matrix element is then defined as the projection of this state on the final state,

$$\langle\Phi_f|.$$

Thus,

$$S_{fi} \equiv \lim_{t\to+\infty} \langle\Phi_f|\Psi(t)\rangle \equiv \langle\Phi_f|S|\Phi_i\rangle,$$

where S is the S-operator. The great advantage of this definition is that the time-evolution operator U evolving a state in the interaction picture is formally known,

$$U(t,t_0) = Te^{-i\int_{t_0}^{t} d\tau V(\tau)},$$

where T denotes the time-ordered product. Expressed in this operator,

$$S_{fi} = \lim_{t_2\to+\infty}\lim_{t_1\to-\infty}\langle\Phi_f\,|U(t_2,t_1)|\Phi_i\rangle,$$

from which,

$$S = U(\infty,-\infty)$$

Expanding using the knowledge about U, gives a Dyson series,

$$S = \sum_{n=0}^{\infty}\frac{(-i)^n}{n!}\int_{-\infty}^{\infty} dt_1\cdots\int_{-\infty}^{\infty} dt_n T\left[V(t_1)\cdots V(t_n)\right],$$

or, if V comes as a Hamiltonian density,

$$S = \sum_{n=0}^{\infty}\frac{(-i)^n}{n!}\int_{-\infty}^{\infty} dx_1^4\cdots\int_{-\infty}^{\infty} dx_n^4 T\left[\mathcal{H}(t_1)\cdots\mathcal{H}(t_n)\right].$$

Being a special type of time-evolution operator, S is unitary. For any initial state and any final state one finds,

$$S_{fi} = \langle\Phi_f\,|\,S\,|\,\Phi_i\rangle = \left\langle\Phi_f\left|\sum_{n=0}^{\infty}\frac{(-i)^n}{n!}\int_{-\infty}^{\infty} dx_1^4\cdots\int_{-\infty}^{\infty} dx_n^4 T\left[\mathcal{H}(t_1)\cdots\mathcal{H}(t_n)\right]\right|\Phi_i\right\rangle.$$

This approach is somewhat naîve in that potential problems are swept under the carpet. This is intentional.

In and Out States

Here a slightly more rigorous approach is taken in order to address potential problems that were disregarded in the interaction picture approach of above. The final outcome is, of course, the same as when taking the quicker route. For this, the notions of in and out states are needed. These will be developed in two ways, from vacua, and from free particle states. Needless to say, the two approaches are equivalent, but they illuminate matters from different angles.

From Vacua

If *a(k)*† is a creation operator, its hermitian adjoint is an annihilation operator and destroys the vacuum,

$$a(k)|*,0\rangle = 0.$$

In Dirac notation, define,

$$|*,0\rangle$$

as a vacuum quantum state, i.e. a state without real particles. The asterisk signifies that not all vacua are necessarily equal, and certainly not equal to the Hilbert space zero state o. All vacuum states are assumed Poincaré invariant, invariance under translations, rotations and boosts, formally,

$$P^{\mu}|*,0\rangle = |*,0\rangle, \quad M^{\mu\nu}|*,0\rangle = |*,0\rangle,$$

where P^{μ} is the generator of translation in space and time, and $M^{\mu\nu}$ is the generator of Lorentz transformations. Thus the description of the vacuum is independent of the frame of reference. Associated to the in and out states to be defined are the in and out field operators (aka fields) Φ_{i} and Φ_{o}. Attention is here focused to the simplest case, that of a scalar theory in order to exemplify with the least possible cluttering of the notation. The in and out fields satisfy,

$$(\square^2 + m^2)\phi_{i,o}(x) = 0,$$

the free Klein–Gordon equation. These fields are postulated to have the same equal time commutation relations (ETCR) as the free fields,

$$\begin{aligned}
&[\phi_{i,o}(x), \pi_{i,o}(y)]_{x_0=y_0} = i\delta(\mathbf{x}-\mathbf{y}),\\
&[\phi_{i,o}(x), \phi_{i,o}(y)]_{x_0=y_0} = [\pi_{i,o}(x), \pi_{i,o}(y)]_{x_0=y_0} = 0
\end{aligned}$$

where $\pi_{\mathrm{i,j}}$ is the field canonically conjugate to $\Phi_{\mathrm{i,j}}$. Associated to the in and out fields are two sets of creation and annihilation operators, $a_{\mathrm{i}}(k)^{\dagger}$ and $a_{\mathrm{f}}(k)^{\dagger}$, acting in the *same* Hilbert space, on two *distinct* complete sets (Fock spaces; initial space i, final space f). These operators satisfy the usual commutation rules,

$$\begin{aligned}
&[a_{i,o}(\mathbf{p}), a^{\dagger}_{i,o}(\mathbf{p}')] = i\delta(\mathbf{p}-\mathbf{p}'),\\
&[a_{i,o}(\mathbf{p}), a_{i,o}(\mathbf{p}')] = [a^{\dagger}_{i,o}(\mathbf{p}), a^{\dagger}_{i,o}(\mathbf{p}')] = 0.
\end{aligned}$$

The action of the creation operators on their respective vacua and states with a finite number of particles in the in and out states is given by:

$$\begin{aligned}
&|i, k_1 \ldots k_n\rangle = a_i^{\dagger}(k_1)\cdots a_i^{\dagger}(k_n)|i,0\rangle,\\
&|f, p_1 \ldots p_n\rangle = a_f^{\dagger}(p_1)\cdots a_f^{\dagger}(p_n)|f,0\rangle,
\end{aligned}$$

where issues of normalization have been ignored. The initial and final spaces are defined by:

$$\mathcal{H}_i = \mathrm{span}\{|i, k_1 \ldots k_n\rangle = a_i^{\dagger}(k_1)\cdots a_i^{\dagger}(k_n)|i,0\rangle\},$$

$$\mathcal{H}_f = \text{span}\{|f, p_1 \dots p_n\rangle = a_f^\dagger(p_1)\cdots a_f^\dagger(p_n)|f,0\rangle\}.$$

The asymptotic states are assumed to have well defined Poincaré transformation properties, i.e. they are assumed to transform as a direct product of one-particle states. This is a characteristic of a non-interacting field. From this follows that the asymptotic states are all eigenstates of the momentum operator P^μ,

$$P^\mu|i, k_1 \dots k_m\rangle = k_1^\mu + \cdots + k_m^\mu|i, k_1 \dots k_m\rangle, \quad P^\mu|f, p_1 \dots p_n\rangle = p_1^\mu + \cdots + p_n^\mu|f, p_1 \dots p_n\rangle.$$

In particular, they are eigenstates of the full Hamiltonian,

$$H = P^0.$$

The vacuum is usually postulated to be stable and unique,

$$|i,0\rangle = |f,0\rangle = |*,0\rangle \equiv |0\rangle.$$

The interaction is assumed adiabatically turned on and off.

Heisenberg Picture

The Heisenberg picture is employed henceforth. In this picture, the states are time-independent. A Heisenberg state vector thus represents the complete spacetime history of a system of particles. The labeling of the in and out states refers to the asymptotic appearance. A state $\Psi_{\alpha,\text{ in}}$ is characterized by that as $t\to-\infty$ the particle content is that represented collectively by α. Likewise, a state $\Psi_{\beta,\text{ out}}$ will have the particle content represented by β for $t\to+\infty$. Using the assumption that the in and out states, as well as the interacting states, inhabit the same Hilbert space and assuming completeness of the normalized in and out states (postulate of asymptotic completeness), the initial states can be expanded in a basis of final states (or vice versa). The explicit expression is given later after more notation and terminology has been introduced. The expansion coefficients are precisely the S-matrix elements to be defined below.

While the state vectors are constant in time in the Heisenberg picture, the physical states they represent are *not*. If a system is found to be in a state Ψ at time t = 0, then it will be found in the state $U(\tau)\Psi = e^{-iH\tau}\Psi$ at time $t = \tau$. This is not (necessarily) the same Heisenberg state vector, but it is an *equivalent* state vector, meaning that it will, upon measurement, be found to be one of the final states from the expansion with nonzero coefficient. Letting τ vary one sees that the observed Ψ (not measured) is indeed the Schrödinger picture state vector. By repeating the measurement sufficiently many times and averaging, one may say that the *same* state vector is indeed found at time $t = \tau$ as at time t = 0. This reflects the expansion above of an in state into out states.

From Free Particle States

For this viewpoint, one should consider how the archetypical scattering experiment is performed. The initial particles are prepared in well defined states where they are so far apart that they don't interact. They are somehow made to interact, and the final particles are registered when they are so far apart that they have ceased to interact. The idea is to look for states in the Heisenberg picture that in the distant past had the appearance of free particle states. This will be the in states. Likewise, an out state will be a state that in the distant future has the appearance of a free particle state.

A general non-interacting multi-particle state is given by,

$$\Psi_{p_1\sigma_1 n_1;p_2\sigma_2 n_2;\cdots},$$

where,

- p is momentum,
- σ is spin z-component or, in the massless case, helicity,
- n is particle species.

These states are normalized as,

$$\left(\Psi_{p_{1'}\sigma_{1'}n_{1'};p_{2'}\sigma_{2'}n_{2'};\cdots},\Psi_{p_1\sigma_1 n_1;p_2\sigma_2 n_2;\cdots}\right)=\delta^3(\mathbf{p}_{1'}-\mathbf{p}_1)\delta_{\sigma_{1'}\sigma_1}\delta_{n_{1'}n_1}\delta^3(\mathbf{p}_{2'}-\mathbf{p}_2)\delta_{\sigma_{2'}\sigma_2}\delta_{n_{2'}n_2}\cdots \quad \pm \text{ permutations.}$$

Permutations work as such; if $s \in S_k$ is a permutation of k objects (for a k-particle state) such that,

$$n'_{s(i)}=n_i, \quad 1\le i\le k,$$

then a nonzero term results. The sign is plus unless s involves an odd number of fermion transpositions, in which case it is minus. The notation is usually abbreviated letting one Greek letter stand for the whole collection describing the state. In abbreviated form the normalization becomes,

$$\left(\Psi_{\alpha'},\Psi_\alpha\right)=\delta(\alpha'-\alpha)$$

When integrating over free-particle states one writes in this notation,

$$\int d\alpha\cdots\equiv\sum_{n_1\sigma_1 n_2\sigma_2\cdots}\int d^3p_1 d^3p_2\cdots,$$

where the sum includes only terms such that no two terms are equal modulo a permutation of the particle type indices. The sets of states sought for are supposed to be complete. This is expressed as:

$$\Psi=\int d\alpha\Psi_\alpha\left(\Psi_\alpha,\Psi\right),$$

which could be paraphrased as:

$$\int d\alpha |\Psi_\alpha\rangle\langle\Psi_\alpha| = 1,$$

where for each fixed α, the right hand side is a projection operator onto the state α. Under an inhomogeneous Lorentz transformation (Λ, a), the field transforms according to the rule,

$$U(\Lambda,a)\Psi_{p_1\sigma_1 n_1;p_2\sigma_2 n_2\cdots} = e^{-ia_\mu((\Lambda p_1)^\mu+(\Lambda p_2)^\mu+\cdots)}\sqrt{\frac{(\Lambda p_1)^0(\Lambda p_2)^0\cdots}{p_1^0 p_2^0\cdots}}\sum_{\sigma_1'\sigma_2'\cdots} D^{(j_1)}_{\sigma_1'\sigma_1}(W(\Lambda,p_1))D^{(j_2)}_{\sigma_2'\sigma_2}(W(\Lambda,p_2))\cdots\Psi_{\Lambda p_1\sigma_1' n_1;\Lambda p_2\sigma_2' n_2\cdots},$$

where $W(\Lambda, p)$ is the Wigner rotation and $D^{(j)}$ is the $(2j + 1)$-dimensional representation of SO(3). By putting $\Lambda = 1$, $a = (\tau, 0, 0, 0)$, for which U is $\exp(iH\tau)$, in,

$$U(\Lambda,a)\Psi_{p_1\sigma_1 n_1;p_2\sigma_2 n_2\cdots} = e^{-ia_\mu((\Lambda p_1)^\mu+(\Lambda p_2)^\mu+\cdots)}\sqrt{\frac{(\Lambda p_1)^0(\Lambda p_2)^0\cdots}{p_1^0 p_2^0\cdots}}\sum_{\sigma_1'\sigma_2'\cdots} D^{(j_1)}_{\sigma_1'\sigma_1}(W(\Lambda,p_1))D^{(j_2)}_{\sigma_2'\sigma_2}(W(\Lambda,p_2))\cdots\Psi_{\Lambda p_1\sigma_1' n_1;\Lambda p_2\sigma_2' n_2\cdots},$$

it immediately follows that,

$$H\Psi = E_\alpha\Psi, \quad E_\alpha = p_1^0 + p_2^0 + \cdots,$$

so the in and out states sough after are eigenstates of the full Hamiltonian that are necessarily non-interacting due to the absence of mixed particle energy terms. The above suggests that the in states Ψ^+ and the out states Ψ^- should be such that,

$$e^{-iH\tau}\int d\alpha g(\alpha)\Psi_\alpha^\pm = \int d\alpha e^{-iE_\alpha\tau} g(\alpha)\Psi_\alpha^\pm$$

for large positive and negative τ has the appearance of the corresponding package, represented by g, of free-particle states, g assumed smooth and suitably localized in momentum. Wave packages are necessary, else the time evolution will yield only a phase factor indicating free particles, which cannot be the case. The right hand side follows from that the in and out states are eigenstates of the Hamiltonian per above. To formalize this requirement, assume that the full Hamiltonian H can be divided into two terms, a free-particle Hamiltonian H_0 and an interaction V, $H = H_0 + V$ such that the eigenstates Φ_γ of H_0 have the same appearance as the in- and out-states with respect to normalization and Lorentz transformation properties,

$$H_0\Phi_\alpha = E_\alpha\Phi_\alpha,$$
$$(\Phi_{\alpha'},\Phi_\alpha) = \delta(\alpha'-\alpha)$$

The in and out states are defined as eigenstates of the full Hamiltonian,

$$H\Psi_\alpha^\pm = E_\alpha\Psi_\alpha^\pm,$$

Satisfying,

$$e^{-iH\tau}\int d\alpha g(\alpha)\Psi_\alpha^\pm \to e^{-iH_0\tau}\int d\alpha g(\alpha)\Phi_\alpha.$$

for $\tau \to -\infty$ or $\tau \to +\infty$ respectively. Define,

$$\Omega(\tau) \equiv e^{+iH\tau} e^{-iH_0\tau},$$

then,

$$\Psi^{\pm}_{\alpha} = \Omega(\mp\infty)\Phi_{\alpha}.$$

This last expression will work only using wave packages.From these definitions follow that the in and out states are normalized in the same way as the free-particle states,

$$(\Psi^{+}_{\beta}, \Psi^{+}_{\alpha}) = (\Phi_{\beta}, \Phi_{\alpha}) = (\Psi^{-}_{\beta}, \Psi^{-}_{\alpha}) = \delta(\beta - \alpha),$$

and the three sets are unitarily equivalent. Now rewrite the eigenvalue equation,

$$(E_{\alpha} - H_0 \pm i\epsilon)\Psi^{\pm}_{\alpha} = \pm i\epsilon\Psi^{\pm}_{\alpha} + V\Psi^{\pm}_{\alpha},$$

where the $\pm i\varepsilon$ terms has been added to make the operator on the LHS invertible. Since the in and out states reduce to the free-particle states for $V \to 0$, put,

$$i\epsilon\Psi^{\pm}_{\alpha} = i\epsilon\Phi_{\alpha}$$

on the RHS to obtain,

$$\Psi^{\pm}_{\alpha} = \Phi_{\alpha} + (E_{\alpha} - H_0 \pm i\epsilon)^{-1} V\Psi^{\pm}_{\alpha}.$$

Then use the completeness of the free-particle states,

$$V\Psi^{\pm}_{\alpha} = \int d\beta (\Phi_{\beta}, V\Psi^{\pm}_{\alpha})\Phi_{\beta} \equiv \int d\beta T^{\pm}_{\beta\alpha}\Phi_{\beta},$$

to finally obtain,

$$\Psi^{\pm}_{\alpha} = \Phi_{\alpha} + \int d\beta \frac{T^{\pm}_{\beta\alpha}\Phi_{\beta}}{E_{\alpha} - E_{\beta} \pm i\epsilon}.$$

Here H_0 has been replaced by its eigenvalue on the free-particle states. This is the Lippmann–Schwinger equation.

In States Expressed as Out States

The initial states can be expanded in a basis of final states (or vice versa). Using the completeness relation,

$$\Psi^{-}_{\alpha} = \int d\beta (\Psi^{+}_{\beta}, \Psi^{-}_{\alpha})\Psi^{+}_{\beta} = \int d\beta |\Psi^{+}_{\beta}\rangle\langle\Psi^{+}_{\beta}|\Psi^{-}_{\alpha}\rangle = \sum_{n_1\sigma_1 n_2\sigma_2\cdots} \int d^3p_1 d^3p_2 \cdots (\Psi^{+}_{\beta}, \Psi^{-}_{\alpha})\Psi^{+}_{\beta},$$

$$\Psi^{-}_{\alpha} = |i, k_1 \ldots k_n\rangle = C_0 |f, 0\rangle + \sum_{m=1}^{\infty} \int d^4p_1 \ldots d^4p_m C_m(p_1 \ldots p_m) |f, p_1 \ldots p_m\rangle,$$

where $|C_m|^2$ is the probability that the interaction transforms,

$$\left|i,k_1\ldots k_n\right\rangle = \Psi_\alpha^-$$

into,

$$\left|f,p_1\ldots p_m\right\rangle = \Psi_\beta^+.$$

By the ordinary rules of quantum mechanics,

$$C_m(p_1\cdots p_m) = \langle f,p_1\ldots p_m\,|\,i,k_1\ldots k_n\rangle = (\Psi_\beta^+,\Psi_\alpha^-)$$

and one may write,

$$\left|i,k_1\ldots k_n\right\rangle = C_0\left|f,0\right\rangle + \sum_{m=1}^{\infty}\int d^4p_1\ldots d^4p_m\left|f,p_1\ldots p_m\right\rangle\!\left\langle f,p_1\ldots p_m\,|\,i,k_1\ldots k_n\right\rangle .$$

The expansion coefficients are precisely the S-matrix elements.

References

- Landau, L. D.; Lifshitz, E. M. (2003). Quantum Mechanics: Non-relativistic Theory. Amsterdam: Butterworth-Heinemann. ISBN 0-7506-3539-8
- Scattering, teaching, ydchong: spms.ntu.edu.sg, Retrieved 13 February, 2020
- Tony Philips (November 2001). "Finite-dimensional Feynman Diagrams". What's New In Math. American Mathematical Society. Retrieved 2007-10-23
- Node, notes: quantummechanics.ucsd.edu, Retrieved 14 March, 2020
- Sakurai, J.J.; Napolitano, J (2011) [1964]. Modern quantum mechanics (2nd ed.). Addison Wesley. Chapter 6. ISBN 978-0-8053-8291-4

Permissions

All chapters in this book are published with permission under the Creative Commons Attribution Share Alike License or equivalent. Every chapter published in this book has been scrutinized by our experts. Their significance has been extensively debated. The topics covered herein carry significant information for a comprehensive understanding. They may even be implemented as practical applications or may be referred to as a beginning point for further studies.

We would like to thank the editorial team for lending their expertise to make the book truly unique. They have played a crucial role in the development of this book. Without their invaluable contributions this book wouldn't have been possible. They have made vital efforts to compile up to date information on the varied aspects of this subject to make this book a valuable addition to the collection of many professionals and students.

This book was conceptualized with the vision of imparting up-to-date and integrated information in this field. To ensure the same, a matchless editorial board was set up. Every individual on the board went through rigorous rounds of assessment to prove their worth. After which they invested a large part of their time researching and compiling the most relevant data for our readers.

The editorial board has been involved in producing this book since its inception. They have spent rigorous hours researching and exploring the diverse topics which have resulted in the successful publishing of this book. They have passed on their knowledge of decades through this book. To expedite this challenging task, the publisher supported the team at every step. A small team of assistant editors was also appointed to further simplify the editing procedure and attain best results for the readers.

Apart from the editorial board, the designing team has also invested a significant amount of their time in understanding the subject and creating the most relevant covers. They scrutinized every image to scout for the most suitable representation of the subject and create an appropriate cover for the book.

The publishing team has been an ardent support to the editorial, designing and production team. Their endless efforts to recruit the best for this project, has resulted in the accomplishment of this book. They are a veteran in the field of academics and their pool of knowledge is as vast as their experience in printing. Their expertise and guidance has proved useful at every step. Their uncompromising quality standards have made this book an exceptional effort. Their encouragement from time to time has been an inspiration for everyone.

The publisher and the editorial board hope that this book will prove to be a valuable piece of knowledge for students, practitioners and scholars across the globe.

Index

A
Angular Frequency, 31, 80, 105-106, 113, 130
Angular Momentum, 4, 7-8, 10-11, 13-14, 18-20, 24, 37, 39, 53, 58, 68, 73, 81, 84, 95, 101, 164, 170-172, 184, 186-188, 194-198, 222
Atomic Spectra, 3, 87, 194-195
Atomic Spectroscopy, 8, 101
Axiomatic Approach, 11, 20
Azimuthal Angle, 162

B
Blackbody Spectrum, 86
Bohr Radius, 57-58, 116
Born Rule, 49, 61
Broglie Wave, 45

C
Cartesian Coordinates, 65, 74, 76, 80, 110, 115, 121, 163, 169, 177
Cesium Clock, 24-26
Continuous Spectrum, 142, 164, 167
Copenhagen Interpretation, 18, 21, 34, 39-40, 48, 64, 87, 98, 103-104
Coulomb Potential, 81, 164, 201
Curie Temperature, 184

D
Diffraction Pattern, 16-17, 29, 219
Dirac Equation, 8, 63, 99, 124, 152, 183
Dirac Notation, 52, 171, 230
Dual Dyson Series, 216-217

E
Ehrenfest Theorem, 109
Eigenvalues, 49, 59-60, 67-71, 74, 79, 86, 91, 96, 99, 101, 107, 111, 114, 132, 154, 160, 162, 165, 189-191, 207, 216-217, 221
Electromagnetic Radiation, 1, 5, 10, 15, 24, 86
Electron Spin Resonance, 185, 193, 207
Energy Eigenstates, 83, 111, 120, 153, 206-207
Expectation Value, 42-43, 70, 84, 91, 100, 125, 133-134, 204, 211

F
Fermi Surface, 169
Fock Space, 64

G
Gamma Rays, 1
Giant Magnetoresistive, 193
Gravitational Force, 4

H
Hamiltonian Function, 100, 110
Harmonic Oscillator, 51, 57, 63, 80, 94, 109, 113, 153-154, 156, 163, 166, 169, 200-201, 217-218
Heaviside Step Function, 144, 149
Hermitian Matrix, 96, 111
Hydrogen Atom, 6-7, 57-59, 86, 115-116, 125, 200-201, 211
Hyperfine Structure, 24, 197

K
Kaon, 22
Kinetic Energy, 3, 6-7, 33, 72, 76-77, 82-83, 104, 107-108, 123, 136, 145, 149, 154, 162, 220
Klein Paradox, 152

L
Laplace Operator, 76, 123
Legendre Polynomials, 63, 223
Lorentz Factor, 31, 33

M
Magnetic Dipole Moment, 15, 82, 181-182, 185
Mass Polarization, 77, 117
Matter Wave, 27, 29, 33, 35-36, 104, 143
Monochromatic Light, 2, 15-16

N
Neutrino Mass, 184
Nuclear Magnetic Resonance, 186, 193

O
Orthogonal Wave Functions, 46, 48

P

Particle Wave, 45, 63, 95, 115

Pauli Exclusion Principle, 180-182, 187, 194

Pauli Matrices, 73, 82, 187, 189, 191, 193

Perturbation Theory, 93, 200-203, 205, 208-209, 211-213, 215-218

Photoelectric Effect, 3, 30, 86

Planck Constant, 28, 31-32, 92, 98, 130, 144, 149, 180-181, 186

Potential Energy, 6, 73, 75-77, 79, 81-83, 104-105, 107, 112, 122-123, 127-128, 130-131, 136-137, 143, 157, 203, 218

Probability Density, 45-47, 55, 58, 66, 110, 122, 132-134

Probability Theory, 86, 91

Q

Quantum Electrodynamics, 14, 183, 194, 202

Quantum Field Theory, 8, 14, 49, 59, 62, 64, 89, 93, 98-99, 128, 152, 223-224, 228

Quantum Tunneling, 101, 113, 143, 147

R

Rabi Oscillation, 197

Relativistic Quantum Mechanics, 123, 152, 180

Rigid Rotor, 81, 218

S

Schrodinger Equation, 46, 157

Special Relativity, 31-33, 40-41, 98, 110, 182

Spectral Theorem, 62, 111

Spectral Theory, 88, 91

Spherical Coordinates, 57, 60, 162, 169

Spherically Symmetric Potential, 161

Spin-orbit Coupling, 53, 194, 201

Spin-statistics Theorem, 180, 182, 189

Stark Effect, 201

T

Tensor Product, 53, 64, 90, 175-176, 223

U

Uncertainty Principle, 1, 36-43, 91, 103, 109, 134-136

V

Velocity Vector, 145, 150

W

Wave Packet, 28, 35, 102, 108-110, 133

Z

Zero Energy, 129, 136, 224